RE-WIRING
BUSINESS
Uniting Management and the Web

Tim McEachern

Bob O'Keefe

John Wiley & Sons, Inc.
New York • Chichester • Weinheim • Brisbane • Singapore • Toronto

This text is printed on acid-free paper.

Copyright © 1998 by Tim McEachern and Bob O'Keefe
Published by John Wiley & Sons, Inc.

Library of Congress Cataloging-in-Publication Data:

McEachern, Tim, 1965–
    Re-wiring business : uniting management and the Web / Tim
McEachern and Bob O'Keefe.
        p.   cm.
    Includes index.
    ISBN 0-471-17556-0 (alk. paper)
    1. Business enterprises—Computer networks.   2. World Wide Web
(Information retrieval system)   I. O'Keefe, Bob, 1958–   .
II. Title.
HD30.37.M39   1997
658'.054678—dc21   97-17565
                                                                CIP

10 9 8 7 6 5 4 3 2 1

Acknowledgments

We would like to thank all who have helped us with the thinking and material that has gone into this book. In particular, Dave Winer for his always interesting DaveNet and quote in Chapter 3, Bill Gurley for his insight in Above The Crowd and thoughts for Chapter 3, Bob Metcalfe for helping start this whole Net thing and his law in Chapter 3, Rick Levine of Sun for adding style to substance in Chapter 4. Gina O'Connor contributed to the thinking that went into Chapter 5, Jim Hopper provided the material on the Redbook incident in Chapter 5, Vince Emery provided the material on the Pentium case in Chapter 6, and Eric Ward allowed us to use a press release and provided quotes for Chapter 6.

No book gets published by itself, we owe particular thanks to Jeanne Glasser, our editor at Wiley, and Neil Levine for first recognizing the potential for a book like this.

Contents

Introduction:
Forward Thinking

USING THE WEB

The Operations Manager

A manager at a medical device manufacturer sits in a Monday morning meeting listening to a briefing. Two members of her staff attended a global company conference last week in Atlanta and called to discuss long-term market strategy. Because her group runs operations and quality control (running production as a result of marketing strategies), she was invited to send two people to listen. One of these attendees, a young recent hire, is upset; he is convinced that the conference saw a decreased role for the product line for which they manage operations, with possible withdrawal from the market within five to seven years. Opening the conference, the senior vice president for marketing had hammered home the need to concentrate on core, profitable lines of business with potential for market growth, even at the expense of mature, profitable lines in low-growth markets. The other person, with over 15 years' experience in the company, feels that the point is exaggerated; the company wants to grow and focus more, but no one is going to kill a profitable product line. He'd heard hot air from marketing types before.

After the meeting, the manager sits at her desk and ponders the briefing. Her view is that although profits for the product line have been slim in the United States, the opportunities for expanding business in Asia and other areas are reasonable. The future may be difficult, but the product is unlikely to be killed. She fires up her Web browser, connects to the company's intranet, and does a key word search on her product name and the name given to the conference. She finds two files containing PowerPoint presentations, one from the senior vice president for marketing, another from a consultant hired to analyze future markets. She downloads both, and then flips through the slides on her PC. Both messages make the same points: Product lines with low margins should be reduced so as to focus efforts on lines with more growth potential. But there are global differences in markets, and some markets that are mature in the United States may have the highest potential elsewhere in the world. *Where* products will be marketed is key, not necessarily *which* products. With the future of her product line secure, she relaxes and wonders why she bothered to send two people to the conference.

The Executive Student

A major corporation can no longer afford the time for their managerial staff to take executive courses. The dilemma is simple—10 years or so after their MBAs, executives need to have their skills "topped up," as a CEO put it. But these are the people who don't have time to go back to some local university for days or even weeks. Moreover, they're distributed around the world, and the last thing any company needs is 50 or so executives getting 50 different perspectives from 50 different academics.

The corporation commits to a single business school that can provide global virtual training. The executives come together for an extensive one-week program in Europe, but this is used more to build teams and generate the requirements for subsequent content, rather than to deliver actual content. For the next year, a series of modules (many requiring teamwork) are delivered via the Web.

While flying back from Europe, one individual's flight is delayed at Heathrow Airport for four hours. He dials the corporation's local service number from his personal computer to collect his E-mail. He has an E-mail stating that a new education module is ready, so he points his Web browser at the course page and downloads the material. After reading some basic material, and visiting the Web sites of some companies that are examples in this material, he downloads further pages that contain a business case. To take part in the case discussion, he connects to the course's newsgroup server; he finds that one colleague has already made some comments on the case (based on typically thin analysis) and takes pleasure in posting a sharp reply. His flight is called; he disconnects and turns off his computer.

The Woodworker's Catalog

A woodworker in Vermont sells her products through a virtual catalog. She makes decorative wooden plates out of unusual woods at the rate of about one per week. Her previous sales outlet has been craft fairs, but these involve a large time and travel commitment. A physical catalog is not an easy option due to the uniqueness of each plate and the expense involved in printing and mailing the catalog.

The virtual catalog contains a picture of each of the plates presently in inventory, plus price details and an order form. She also lists her phone number. When a plate sells, it is removed from the catalog; when she finishes a new one, she adds it to the catalog. At any time, approximately 20 to 30 plates are available. The virtual catalog is hosted by a local Internet Service Provider (ISP). She pays $15 a month for the account, plus another $50 a month for them to do the maintenance work on the catalog. She E-mails them necessary changes with digital pictures as E-mail attachments, when needed. She takes these pictures using a $400 digital camera and then copies the file onto her PC before attaching it to an E-mail.

Every day after working in her workshop, she dials up her account and downloads completed orders and E-mails. Immediate completed orders from visitors to her virtual catalog are very rare; more likely she receives E-mails from people interested in her plates, wanting to know more about the wood, the sizes, the process. She often has long E-mail exchanges with potential customers. Some never buy anything, but some do buy. She knows she has a sale when someone actually calls her—it's as if they just have to check that she really does exist.

As part of her E-mail experience, she has had to become adept at responding to varied questions and requests, many posed in rather broken English. But she has learned to persevere—E-mails from Japan, for example, are more often converted into sales than E-mails from Australia. Half of her products are now shipped overseas.

Recently, her business has taken an interesting turn. People suggest designs and woods to her. She then creates the pattern and makes the plate. It's hardly what you'd call customized design since there's no guarantee that the potential buyer will purchase, but she's found out that if someone wants something in particular and it makes sense, then it's a plate that someone somewhere will buy.

The Consultant's Printer

It is Tuesday afternoon in upstate New York. A business consultant, who sells his skills to a number of corporations, is working on a proposal for a potential client in Japan. He has to complete the proposal by Wednesday afternoon. He'd actually have until Friday morning if the person requesting it would accept an E-mail attachment, but this individual wants it sent the old-fashioned way. Thus, it will spend nearly two days being physically transported halfway around the world.

While printing out a draft, tragedy occurs. The laser printer dies. It doesn't just jam, it actually *dies*. There's even some smoke. Our consultant could drive 30 minutes to the nearest computer superstore, but leaving his workplace isn't convenient at this time.

The consultant fires up his Web browser (he's *always* connected to his Internet Service Provider (ISP) when working so he can quickly respond to E-mail). There are three or four virtual stores that should stock exactly the configuration of printer he wants, and he chooses the URL of one of them from his bookmarks. It has the model, at a great price, but it is not guaranteed to be in stock. He jumps to another store and finds the model (priced a bit higher) but in stock. They can guarantee delivery to his home by midday tomorrow. He places an order; he pays for it by typing in a credit card number (which is suitably encrypted, of course).

On receiving the order, the virtual store's server passes the details to its logistics company, which holds all the store's stock packed and ready for shipping; the virtual store holds no physical inventory at its place of business. (Identifying where its place of business is, in fact, rather difficult, since the server is actually run by the store's ISP.) The payment details have already been forwarded to the appropriate credit card company. The logistic company's system schedules the package for overnight flight and gets it retrieved from the automated warehouse. It then bills the virtual store for the operation.

At 6 A.M. the next day, the package arrives in Rochester, New York. It is routed to a delivery van that starts its first route at 8 A.M. At 10:30 A.M., the consultant's front doorbell rings. At 11 A.M., he prints his draft.

INTERNETWORKING, INTRANETWORKING, AND THE WEB

These examples are not views of the future. They reflect scenarios that can, and are, happening today. The technologies that tie the four scenarios together are *Internetworking* and *intranetworking*, the ability to be able to tie computers and people together over public networks (the Internet), private networks (intranets), or some combination of the two. The set of protocols and software that makes the networking come alive is the World Wide Web (Web), probably the most talked about technology of the decade, and certainly the most written about.

To marketing people, the Web is a global marketplace (or what some call *marketspace*). To Information Systems (IS) executives, it's a cure for many corporations struggling to develop internal information and communications systems (via intranets). To entrepreneurs, it's a market of millions of potential customers for services and products. And to Wall Street, the Web has been a source of outlandish initial public offerings (IPOs) that have seen companies ranging from technology providers like Netscape to services like the Web directory Yahoo! valued at levels far beyond any reasonable multiple of their earnings.

If the Web is the technology of the decade, then *virtual* is the word, and *cyber-* is the prefix. We have virtual work, the virtual corporation, virtual malls, and virtual communities. We have cyberspace, cybernauts, and even cybercafes.

Journalists have written more words about the Net and the Web than they wrote about the Gulf War. They have attached an almost mythical meaning to the Web, and reported it as a conduit for advertising, marketing, publishing, pornography, or junk mail as convenient. But, from a business perspective, the Web is really nothing more than a technology *enabler*. It enables us to do things differently, perhaps radically. It's the *thing* we do, however, that makes business sense, not necessarily the way it's enabled. Admittedly, the networked world now being created will result in new businesses as yet unthought of, and the Web is part of this. But just having a Web server won't make anyone rich. (Developing the software for a Web server or browser has already made those involved in some IPOs rich, but that's a different story.)

The Operations Manager gives a glimpse into one of the roles of intranetworking—the ability to publish presentations, white papers, regulations, contracts, and other documents within an organization, and have anyone quickly find them. Information Systems have traditionally been good at moving information up through organizations, with transaction data being summarized for management.

But the distribution of knowledge—plans, ideas, concepts—has in many cases been terrible. Intranets have been a revelation, allowing for the horizontal flow of information between functional areas, business units, and even vendors and suppliers (networks that some call extranets).

The Executive Student is a less obvious story. But it is perhaps the most important. Many people, not only business executives, spend much of their working life at institutions that give them knowledge. But we now have a medium for effectively transporting knowledge to them, anywhere in the world, at any time. And students can collaborate through groupware both to analyze the knowledge and to apply it. Education, particularly for those short of valuable time, should be a pull action that allows students to obtain it as needed; too often it is a packaged activity delivered at the convenience of colleges and faculty.

The Woodworker is using the Web to reach a global market through a dynamic medium. But the major factor is her ability to deal with potential customers and pleasantly close sales. The story has, perhaps, more in common with real estate and used cars than, say, producing software. Moreover, her product is changing from premade plates to customized work: By interacting with her market to her fullest extent, she learns about her market. This is straight out of the first lecture from any course in entrepreneurship. You don't learn about a market without listening to it.

In the Consultant's Printer, the Web is only part of the story. The integration of virtual retail with logistics is what allows our consultant to get his printer. The virtual store is providing a service—guaranteed delivery within a time period. The management of the product is secondary and is managed by another entity. The virtual store is created from the marriage of two technological advances—distribution logistics and Internetworking.

The Networked Organization and Customization

Another way of perceiving these scenarios, and other changes in business, is to realize that traditional organizational structures are being blown apart. We are moving to a world of networked organizations that dynamically link workers, knowledge, and customers. Ives and Javenpaa[1] paint a future of "knowledge nodes" that can be tapped to provide customized services that can then be combined and delivered anywhere in the world, perhaps electronically. A business consultant, for example, can search for and combine industry-specific expertise, financial analysis, and even the skills to produce a report and a presentation in a very short time and then deliver the combined results electronically.

A key aspect of the future is *customization:* the ability to take a generic product or service and customize it to fit the individual. Lawyers (think house closings), accountants (think tax returns), and home builders have been doing this forever. What is different now is that networking provides an environment that encourages customization. Since the knowledge and services being provided to the customer can be dynamically altered without having to rebuild physical structures or tap into different physical distribution channels, the natural tendency is to move from an a la carte operation to one that dynamically alters the service or product to fit the needs of individual customers. This is the path being taken by our Woodworker. The business school providing virtual training (or *distance learning*) is likely involved in customization by using some of its core material for all programs, and then developing new material and hiring adjunct teachers to provide additional content for a customized program.

Think how much more difficult it is to arrange these customized programs in the physical world, where content, teachers, and students must be brought together in one place and at the same time. The Web is not a necessary condition for customization, but it is a strong enabler.

Outsourcing

In business-speak, companies that add value by customization often retain core competencies and outsource noncore operations. What does this mean? Basically, if you're selling customized T-shirts on the Web (i.e., you work with the customer to create the design), then your core competencies are design and customer interaction. You don't produce the actual T-shirts, and you don't run the Web server. You also probably hire an accountant, lawyer, and other professionals when necessary.

Luckily, many entrepreneurs can get hip to this without understanding the lingo. It is unlikely that the explosion of entrepreneurship on the Web could have happened without the recent outsourcing trend in the past decade or so, and it is certain that it could not have happened so quickly. This is because many companies, ranging from logistics to networking to producing T-shirts, are lining up to sell their service to other companies. Fifteen years ago, a small direct sales operation would often package shipments and send someone down to the post office to stand in line. Now, one of a number of companies will provide you with packaging, pick up the shipment, guarantee delivery within a specified time frame, and even let you track the package using the Web and their computer systems. And all at *variable* cost, allowing a start-up enterprise to focus on marginal revenue and costs.

Culture and Globalization

Jacques Chirac, president of the French Republic, referred to the Web as "American imperialism." The leadership in protocols, technology, and content provided by the United States is not something that every culture is trying to emulate. The notion that the rest of the world is trying to catch up with the United States is pure arrogance, particularly in France, where the Minitel system provided many of the information-sharing benefits of the Net well before it was the behemoth of today. The British invented the modern postal system and hence don't have to put the name of their country on their stamps, but this hardly equates to a national competitive advantage for them. Other countries adopting the postal system have changed and improved it as appropriate. Similarly, Internet users in the United States don't need to use a country code but may notice that other countries adopt or change the structure and usage of the Net in their country to suit local requirements and cultural predilections.

The globalization of business was evident well before the popularity of the Net, and would continue even if the world pulled the plug on the Net tomorrow. But like customization, the Net enables and accelerates the process. Cultural differences in communication (let alone language differences), varying regulatory environments, and societal attitude to electronic commerce are just three of the things that business has to contend with. Want to know why an E-mail across an intranet to England didn't get a response? Not because it didn't get through, but because the British tend to see E-mail more as a written communication medium that requires thought and a reasoned response, whereas Americans generally think of it as a written phone system.

In our scenarios, the Woodworker takes advantage of globalization by selling to overseas customers and the

Executive Student is enrolled in a globally available program. Less obvious, why is the Consultant purchasing his printer from a virtual store in the United States? Surely retailing can now move to where the printers are actually assembled? The value chain can collapse even further. Understanding cultural differences and knowing where we presently stand in the process of globalization are essential.

Technological Choices

If you have read to this point with great suspicion and have questioned the need for the Web in *any* of the stories (let alone all four), then please congratulate yourself. Variations of each story could happen without the Internet and the Web being anywhere in sight. The Consultant could phone around various companies that will guarantee next-day delivery; many computer and peripheral manufacturers provide toll-free numbers for orders. Or he may have physical catalogs at hand.

The Woodworker doesn't have to sell directly, and the effort is obviously considerable. Successful catalogs specializing in craftwork abound. For someone wanting to be free of the time and travel commitment of craft fairs, E-mail seems to be a replacement burden. The Operations Manager, may have a tailor-made *Executive Information System* that provides her with all the market strategy information she needs, or may just make a few phone calls.

Perhaps most obviously, the Executive Student may be better served by a proprietary solution such as Lotus Notes. Better security, easier off-line working, and more structured newsgroups are just three of the present advantages.

While there also are excellent arguments for the Web being the first-choice technology in each case (which we won't dwell on now), the important fact is that the Web is no more than a technology *choice*. It is not a panacea. Business finds itself in a double dilemma. Not only are the Internet and the Web allowing things to be done differently—and business needs to work out what these *things* are—but solutions may involve technologies other than, or in addition to, the Web.

Anyone who tries to sell a solution based solely on Web technology without consideration of alternative technologies is probably a fool. For many companies, Lotus Notes as a packaged solution to group work, document management, and simple work flow is more sensible than the Web. *Electronic data interchange (EDI)*, used by companies to exchange transactions through third-party networks, has not been replaced by the Web (although EDI's growth has certainly diminished since the advent of the Web and it will probably undergo radical changes). And video on demand through your local cable company may be a better bet for future domination of America's Saturday nights than video through the Web.

Despite this caveat, the Web will continue to be a dominant technology. More importantly, the Web as we know it today is really only a platform for applications. It is global client/server computing where any PC is a client, and any company can provide a globally accessible server. With changes now taking place, this platform can be used to deliver a multitude of applications. The difficult part is envisioning these applications, not necessarily making them happen.

FORWARD THINKING

So why are we all here? This book is an exploration of the business advantage that can be gained from the Web. It's about Web marketing, and intranets, and changes in whole businesses. How managers need to rethink the way they can do business and reorganize their departments and companies to accomplish new goals. But it's mainly about *how* the various pieces fit together, and the extent to which they can (explicitly or implicitly) support the ability of a business to function, grow, survive, make money, or just plain have some fun.

1

Liftoff

In thinking about the Web and enterprise, where do you start? Starting at the beginning is always a good idea. Early adopters often make all the mistakes, but sometimes 'reap most of the advantages. Despite the youth of commercial use of the Web, a few benchmarks stand out—companies that have accrued a sustained benefit from their Web efforts. In this chapter, we look at two operations that are about as far apart in the business world as you can find: a start-up retail operation owned and operated by one person, and a division of what is now the world's largest conglomerate.

NINE LIVES CLOTHING STORE

Nine Lives Clothing Store was opened in February 1993 by Mary Jane (MJ) Nesbitt in Los Gatos, California. MJ had been pursuing a legal career, but had always wanted to open up her own enterprise. In 1993, she abandoned caution and created Nine Lives.

Nine Lives is a clothing store where customers buy preowned clothing and contract with Nine Lives to sell their unwanted clothes (the term "used" seems to have been banished in this area of commerce). What makes Nine Lives different from other consignment stores is that it deals exclusively with high-quality clothing from name designers—Anne Klein, Liz Claiborne, Donna Karan, Eddie Bauer, among others. Items are sold for approximately one-third of the original price, and Nine Lives splits the proceeds 50:50 with the consignor. Los Gatos is at the edge of Silicon Valley, which has taken the American notion of "dress for success" to extreme.

In the spring of 1994, Nine Lives became one of the first small businesses on the Web. They were an adopter when there were only a few hundred commercial servers with publicly accessible Web pages, and many were hardware or software companies. MJ's husband, David, established a server on a Unix-based 386 PC with 16 MB of memory and a 14,400 baud modem (by any measure a modest machine). The server was free software from the National Center for Supercomputer Applications (NCSA), the Center that produced the browser Mosaic. Every morning, the computer (or PC as we'll call it, which sits in MJ and David's home) phones the local ISP, makes contact, and then waits for Web users to access it. The store's inventory database shares the PC with the Web server and is accessed from the physical store by an ancient computer and a dial-up connection.

The Nine Lives Web page introduces visitors to the store. They learn how the consignment operation works and can see a picture of the store. But David was not content to have the Nine Lives page simply act as an electronic brochure. A capable C programmer, he interfaced the pages with the store database so that visitors on the Web could query the database, obtaining immediate on-line results to their queries. (See Figure 1–1 for an example.) Users complete a form specifying the size, price, and types of clothing (skirt, dress, blouse, jacket, etc.) they're interested in.

Moreover, David developed the notion of a *personal shopping assistant*. The query generated by a user can be saved as a *personal shopping profile*, which, if so desired, can be retained as a personal shopping assistant. The assistant has a user-generated password attached, and users can access the assistant and cancel or change it as desired. For each assistant, inventory is checked daily, and the customer is E-mailed when the query produces a match.

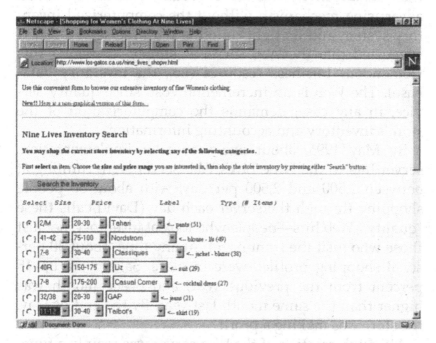

Figure 1–1. Have My Agent Call Yours—
Nine Lives Shopping Assistant.

Following David's work, Nine Lives became a virtual shop as well as a physical shop. Web access in summer 1994 varied from about 10 to 20 hits (visitors) per day, sometimes exceeding walk-ins. The personal shopping assistants reduce the cost of monitoring inventory as it arrives to almost zero, according to David. Further, for a Web user, the effort involved in finding out if a desired item is available may be easier than visiting the store—the traditional information cost for the shopper is lowered. This is convenient for a high-class consignment store since consumers may well have already seen the clothing item they wish to purchase, and just want to check whether they can obtain a less expensive preowned garment.

Despite operating a mom-and-pop store, MJ and David were fully aware of the value of integrating the Web with their database. The Web server would be no more than an advertising mechanism without the computerized inventory system and its tight coupling with the server. David suggested that the Web server is simply leveraging this tremendous business resource (i.e., the inventory database). The Web is an incremental cost, since quality service, in any case, demands the computerization of the store's inventory and accounting information.

By May 1995, about one year after implementation, server hits to the Nine Lives Web server were running at between 1,500 and 2,500 per day, with about 50 people shopping through the server each day (David calls these "quality" Web hits—people who look at inventory, not just those who visit the front page). At any time, about 30 personal shopping profiles were in place. Sales were up 15 percent from the previous year, and every month was higher than the same month last year. By the end of May, the store was making a profit.

Although creation of the Web server was mainly a function of David's enthusiasm, MJ and David now had a full understanding of the target market for their Web service. Like many small businesses, success depends on capturing the local market. David suggested that the Web server is part of a premium service, not a way of directly selling clothes:

We are within driving distance of over 1 million people, so making simple information (hours, location, product description) available is worth the cost of the connection. The other services (inventory lookup, personal shopping assistants) create premium services for on-line browsers. No sales through the Net—the product must be examined before being bought. Despite this focus, sales are occasionally made to people visiting from other areas who have discovered Nine Lives through the Web. A lady from Ireland, after browsing the inventory on the Web, visited the store while accompanying her husband on a business trip. She spent over $200.

In the summer of 1995, David discovered that the personal assistant program had a bug in it, causing over half the attempts to hire an assistant to fail:

The repaired version of software has allowed new personal assistants to be created at the rate of more than one a day. By the end of this year (1995), we project we will have more than 200 active profiles. MJ sends (thanks to the automated searching and mailing capability) between 20 to 50 E-mail messages a day to the owners of the profiles.

In fall 1995, quality Web hits were steady at around 100 to 200 inventory queries a day. In October 1995, the store recorded the best single day of sales, best week, and best month. Profit margin for the month was 25 percent on sales of around $20,000. As of April 1996, the growth in business showed no sign of slowing. The number of phone calls from Web users asking for garments to be held is increasing, and the store has now expanded to include men's clothing (with the necessary change in the database and implementation of the query mechanism).

The personal assistant is a software *agent*, but it was programmed before most people knew what an agent was and certainly before David had heard of the word. Agents are now an important Net technology that allows people to get notice from the Web when a service or product is available instead of having to go back repeatedly and query sites. The Nine Lives personal assistant may be the first ever commercial functioning agent (*not* a prototype) resident on the Web.

GENERAL ELECTRIC PLASTICS

General Electric Plastics (GEP), a division of the world's largest conglomerate, makes the raw plastics used by others to make, well, just about anything. Anybody who has handled the plastic outer cover on a laser printer or caulked a bath has touched GEP products without knowing it. The market for specialized plastics is well developed, dominated by a few players, and global. Key competitors include Du Pont, Ciba-Geigy, and Monsanto.

Summer 1994 on the Net was like summer 1969 to hippies—the little world in which these early participants existed was about to explode. That summer, the general manager of marketing communications at GEP, Rick Pocock, faced a strange dilemma. GEP's products were selling at a record-setting pace but the image of the division was slipping. Competitors had instituted programs such as CD-ROM product catalogs and on-line data sheets that his customers felt showed clear technology leadership. Rick needed something big—something that would cause a splash, not only in the plastics industry but also in the general business press.

GEP's public relations firm, Commsource, had two people with high-tech backgrounds, having worked with several software companies on several product launches. They knew that to break into the general business press with a technology campaign GEP had to have something that no other competitor had. Several brainstorming sessions produced good ideas, from CD-ROMs to CompuServe forums, but Rick felt that none could generate the necessary "buzz." Commsource brought in one of us (Tim) to discuss electronic marketing possibilities. The five-hour meeting that ensued provided plenty of opportunity for brainstorming GEP's various options, from CD-ROM to private on-line services to the Internet. The sheer amount of enthusiasm and energy in the room gave Rick confidence that he would find his hypemaker.

The decision-making process occurred over a two- to three-week period. Various options were researched, presented to the group, and acted on. CD-ROM was the first off the list since the technology was limited to people with CD-ROM players. While this was a growing segment of the PC marketplace, the numbers of PCs with CD-ROM drives installed did not warrant the investment in developing the CD-ROM. It was decided that, once GEP's materials were converted to an electronic form, this option could be revisited.

AOL and Prodigy were quickly dismissed as being too consumer oriented. GEP markets its products and services to other manufacturers; the division is not in the consumer product business. GEnie, the GE Information Services on-line service, was also quickly dismissed as lacking the appropriate subscriber base. At the time, GEnie was primarily aimed at the youth market and the on-line gaming community. This presented a potential PR nightmare—a GE business unit contemplating its first on-line offering and shunning GE's on-line presence. Even with this potential for liability, a GEnie approach was quickly discarded.

CompuServe at the time had the largest subscriber base and a true business focus. Numerous manufacturers had established electronic forums. These were being used as customer support centers, dealer locators, informal customer research groups, and public relations and in some cases for the distribution of business software and software updates. GEP decided to approach CompuServe to create an informational area for the entire plastics industry. CompuServe asked GEP to submit a formal proposal and told GEP that a decision would take approximately four months. This solution did not meet GEP's timetable and the CompuServe option was dropped.

Tim suggested that GEP design a Web site. No Fortune 500 company (or company division) outside the telecommunications or software industry had built a large-scale site, and if done quickly, it would achieve Rick's objectives. GEP and One World created a comprehensive plan to convert over 1,500 print pages of technical material into *Hypertext Markup Language* (HTML)—the input language on the Web. The overall objective that Rick and his small team laid out was simple and ambitious; no fluff, nothing that the Internet community would view as blatant advertising, in short, real data that real engineers could use (see Figure 1–2).

The first task was deciding what material to place online. GEP had recently cataloged its technical library with the help of its local advertising agency, R.T. Blass (the parent company of Commsource). During a meeting that included R.T. Blass, GEP, and One World, the participants discussed what would be appropriate to put on-line, what information might fall into competitors' hands, what levels

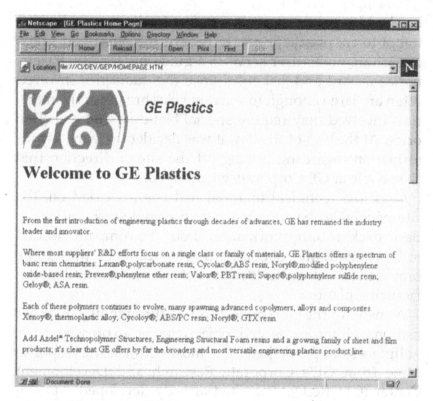

Figure 1–2. GE Plastics Home Page—Circa 1994.

of security would be available, and related issues. A fairly enlightened decision came from the meeting: All the information that GEP would provide would be completely open with no security restrictions and the latest and most technical information that GEP possessed would be placed on-line. This included design guides, injection-molding guides, and product data sheets.

The major dilemma in deciding what would go on the Web centered around the placement of pricing. Since there are a limited number of raw plastic suppliers and the suppliers are all members of the Fortune 500, GEP was worried that providing pricing information might be viewed by the U.S. government as a form of collusion and an attempt to fix plastic prices. In reality, the published prices of each manufacturer serve as a guideline for purchasers, not as a hard-and-fast rule, since the quantities involved often are large enough to warrant price breaks or the materials involved may require special blending and a custom price. At the end of the day, it was decided to leave any information regarding pricing off the site, a direction that the people at GE Corporate encouraged.

Once the technical material had been decided on, the site needed to be rounded out to complete the offering. Basic background information about the company needed to be created, and opportunities for feedback were to be provided as well as the locations of sales offices and manufacturing plants.

As news of the GEP project leaked throughout GE, other factions either presented objections or expressed their willingness to participate. The most enthusiastic response came from GE's Corporate Research and Development Center (GECR&D). They had already developed a prototype Web page for the company and had been offering their own information on-line for a short time. An initial meeting between the two groups proved tense, since each group wanted to be viewed as bringing this technology forward within the company. GECR&D felt that it was well within their charter to champion this effort; however, GEP had the marketing muscle to make the site launch a major event. GEP eventually decided to go ahead with their plans regardless of what the other GE businesses decided to do. GE Corporate supported them and told the other businesses to coordinate their efforts with GEP. At the time of launch, GE Plastics, GE Corporate, GECR&D, and GE Information Services went on-line the same day.

GEP's major concern was how the Internet community would accept their on-line offering. Several members of the team expressed concern over security and *flaming*. As it happened, security became a major issue for GE. The server was initially housed in the GECR&D facility in Schenectady, New York, but a decision was made to move the server to the company's telecommunication facilities in Princeton, New Jersey. Through an oversight at GECR&D, numerous user accounts with passwords were left on the machine. A hacker broke into the machine shortly after the Web launch and went undetected throughout the entire GE internal network until Thanksgiving week, 1994.

The *netiquette* issue (etiquette on the Web) was much harder to resolve and prepare for. For a company with the global presence of GE, deciding to venture into a completely new communications medium was a major risk. Common questions included, "What if we get flamed?" and "What if the Internet community views this as a commercial intrusion and we get negative feedback?" There were no answers to those questions until the medium had been tried. Doubts persisted through the launch and into the postlaunch meetings. This concern was a driving force in determining the initial content of the site. By taking a neutral, unobtrusive and decidedly noncommercialized approach to distributing technical information, it was hoped that the backlash would never materialize. As it turned out, the press feedback, the phone calls, and the E-mails received after the launch suggested that the Net community fully accepted this approach.

Marketing the site launch quickly took on a life of its own. A stake had been set: October 14 would be the firm launch date.

Commsource designed a promotion plan targeted to lure the three categories of press to the launch: the plastics trade press, the computer trade press, and the general business press. To maximize the chance that the press would arrive, GEP made some tactical decisions: The launch would occur in October, a three-stage mailer would be sent with very little information as to what was being announced, free software would be given away (Spry's "Internet-in-a-Box"), and (perhaps most importantly) they would hold the launch in the Rainbow Room at Rockefeller Plaza. The promise of a free meal in a glitzy atmosphere would usually attract even the most jaded reporter to virtually any function, and in this case it proved more than true. CNN and CNBC sent camera crews, and *Newsweek, BusinessWeek,* CBS radio, all the PC and plastics publications—a regular who's who of the press—came, watched, and reported for six straight months. Even today you still see some mentions of the event and GE's role on the Web.

The biggest marketing question facing the GEP team was what if they developed a Web site and their customers weren't on the Web or didn't have access to it? This was a very real problem for Internet pioneers.

At the time, there were no commercially available Web browsers. Netscape had just formed a company, Spyglass was in the process of negotiating their licensing deal with NCSA, and NCSA gave away their browser without a means of connecting to a service provider and or technical support. A relationship with Spry Inc., the makers of Internet-in-a-Box, was formed based on Spry's ability to deliver their announced software by the time of the press conference. They had previously announced that the first version would be ready by summer 1994, but that date slipped to October. (Spry was acquired by CompuServe in early 1995.) Spry contracted with GEP to provide customized versions of software, including a version of Mosaic that featured the GE logo rather than Spry's and a customized *newsgroup* reader that would point to industry-related newsgroups. David Pool, president of Spry, was invited to participate in the press conference in New York City.

The response that GEP received was impressive—over 3,000 hits per day on a technical site from an audience made up of primarily engineers. GEP quickly moved its efforts from the public relations mode to true communications mode, adding response forms, E-mail routing, on-line training, technical tips, and expanded product information updated monthly. While many within the organization still doubted the value of the medium, a highly qualified base of potential customers was gaining easier access to GEP information than to that of any competitors.

Today, the GEP site remains the major benchmark for those wanting to publish technical information on the Web. And, as will be covered in Chapter 3, the site proved to be a model of business-to-business interaction that was extended throughout GE.

SUSTAINED ADVANTAGE?

Nine Lives has been used extensively as a small business case. GE Plastics will forever be the early adopter example of Web technology by a noncomputer Fortune 50 company. But where's the beef? Have these Web sites provided their companies with a sustained competitive advantage? Or is the benefit nothing more than good press (and a mention in this book)?

Open any good book on the mechanics of Web marketing (there are way too many to choose from) and you get a list of the wonderful things that a Web site can do for you. Table 1–1 shows some commonly stated advantages, in part adapted from the book *World Wide Web Marketing* by Jim Sterne.[1]

The Nine Lives Web site certainly generated new customers—including the example given of the visitor from

Table 1–1. The advantages provided by Web to Nine Lives and GE Plastics.

Advantage	Nine Lives	GE Plastics
Generate new customers	Definitely	Maybe
Retain customers	Maybe	Yes
Increase sales	Yes	Maybe
Improve image	Yes	Definitely
Reduce communication costs	Definitely	Yes
Reduce transaction costs	Maybe	No
Increase efficiency	Yes	Yes
Increase profits	Definitely	No

Ireland. Their sales have increased since the creation of their site, and the store has become profitable, but the business may have taken off anyway. Then again, three other consignment stores in the Los Gatos area have opened *and* closed since Nine Lives opened its doors. What is clear is that the shopping assistant technology has reduced the marginal cost of informing customers of new arrivals into inventory to zero.

General Electric Plastics is harder to analyze, as would be expected for such a large operation. Conceived as something that would give GEP a better high-tech image, it has certainly achieved that goal. But the site has become something more: a focus for much of GEP's information management. Whereas technical plastics information, the heart of the business, was kept in paper form and updated infrequently, updates now occur once a week. Customers, the salesforce, and GEP engineers use the Web site to get accurate specs. The publishing of technical information is more efficient and timely. There is also anecdotal evidence that the site has generated new customers—E-mails have been routed to the sales operation, who have followed up inquiries, sometimes with successful results. However, whispers were heard suggesting that such leads may have just been diverted from more traditional channels.

If imitation is the sincerest form of flattery, then the people at GEP should feel well flattered. Most plastics companies now have a site that contains some technical information. However, few have seen fit to publish the level of detail that GEP provides. Moreover, it is not clear that GEP's competitors have been able to integrate E-mail routing and response necessary to make a business-to-business site more than an advertisement. Not only does GEP route E-mail inquiries based on product name, but it uses analysis of site access statistics to help understand what technical information people look for, and how they compare information across products. In addition to all this, GE has made key efforts to stay ahead of the later adopters. Corporate R&D engineers, for example, have experimented with providing plastics simulation models on the Web so that customers not only can look at technical results, but can run their own simulation models to receive specific results. Now that's *engaging* customers.

2

Strategic Matters

What are the business approaches for the successful use of the Web? How can we think about the Web to gain a strategic business advantage? In this chapter, we look at what works and what doesn't when incorporating the capabilities of the Web and the Net into your company's strategic plan.

Our thinking is that there is not one way or one approach, but a growing number of identifiable approaches to either integrating the Web into present business strategies or using it to develop new businesses. In this chapter, we identify these approaches. First, however, it is necessary to identify what the Web actually *is*. Not the protocols, the hardware, or the bandwidth, but how the thing should be viewed from a business perspective. If we think of the Web as, say, electronic publishing, then, we're going to develop a set of Web pages for our business and publish some material that looks the same as our existing physical publications. And that is exactly what many businesses have done. How we perceive something can often be a limitation; here we want it to be the opposite: something that broadens our thinking. The knack is to find the right analogies.

There are at least three perspectives that can bring insight and help frame your thinking about the Web. First, the Web can be perceived as a *marketplace*. It's a place where suppliers can offer (and perhaps even supply) their goods and services, and customers can find and purchase them. The global electronic marketplace is a dream opportunity for marketing executives. What the reality ends up looking like is another matter. But nonetheless, the Web as a market is perhaps the most important perspective, and a far more expansive analogy than electronic publishing.

Second, the Web is a global interorganizational information system (or global IS). Organizations and individuals can communicate across a single system, as long as they follow the rules of the road (i.e., the various Internet protocols). Suddenly, someone sitting in a company with a local E-mail system that only about 100 people actually use regularly can send E-mail to over 30 million other people. It's like having a phone company that routed only local calls getting long distance access. Moreover, companies can share data and use the Web as a front end for transaction processing. We now have a technology that makes EDI look as useful as FORTRAN.

Third (and here's the odd one), the Web is a *cultural* and *global* phenomenon. As an entity in its own right, the Net has generated its own language ranging from little smiley faces that get shoved in E-mails to shorthand mnemonics such as IMHO (In My Humble Opinion). And as discussed in the Introduction, pretty much everything is either cyber or virtual now. The rush to commercialize the Web has not blown away this culture, and in fact, some rather staid companies have made attempts to adopt it. There are considerable implications for how we communicate, and the extent to which older workers can adapt.

This culture, however, is a veneer. Underneath are the far more ingrained cultures of different countries, societies, and organizations. To relate this to the Web, a Web site that appeals to an American audience not only may look stupid elsewhere in the world, but may actually be viewed as insulting. In addition to the consequences for global commerce, this is something that global companies have to face up to when using the Web. In this chapter, we develop these ideas further.

THE WEB AS A MARKETPLACE

Our grandfathers would take their agricultural products to market for sale at a fixed, agreed, or auction price. In the United Kingdom, the four annual market days still represent days that segment the year. The Web is now a global marketplace—open 24 hours a day, 7 days a week.

The launch of a single product into the physical marketplace is something that firms take lightly at their peril. With the Web, companies have to consider the launch of products and services into a different exchange medium. Using some concepts drawn from new product development, we see the Web has a number of *attractors* that tempt, or pull companies to enter the marketplace. Companies also have *strategic drivers* that push them onto the Web. Finally, the Net can be characterized by certain dimensions that *inhibit* entrants, or cause entrants to withdraw. Figure 2–1 shows these three influences (attractors, drivers, inhibitors) and identifies instances of each that we will consider here.

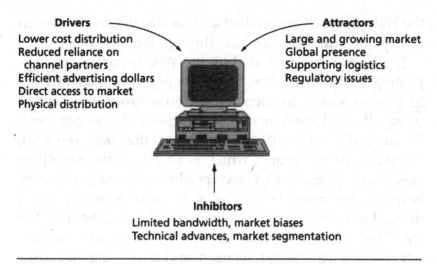

Drivers
Lower cost distribution
Reduced reliance on
 channel partners
Efficient advertising dollars
Direct access to market
Physical distribution

Attractors
Large and growing market
Global presence
Supporting logistics
Regulatory issues

Inhibitors
Limited bandwidth, market biases
Technical advances, market segmentation

Figure 2–1. The Web as a Marketplace.

Attractors

What attracts a company to the Web? There is certainly a large element of "because it's there," but this isn't the only reason. Serious players are attracted by specific features that make entry into the Web appealing. Typically, they are looking to sell a product or provide a service, and thus view the Web as a marketplace that shares many characteristics with retailing.

Large and Growing Market

Whichever way you cut it, there are a lot of people out there. Viewed as a marketplace, the Net offers more than 30 million global consumers. A recent CommerceNet/Nielsen study[1] estimates that more than 2.5 million people have already purchased a product or service through the Web, and the analysis of that data by Hoffman, Kalsbeek, and Novak[2] estimates over 16 million "hard core" Web users—people who use the Web at least once a week for a multitude of tasks. Other studies have consistently found that people use the Web to search for product and service information, even if they don't directly purchase through the medium.

The huge reach provided by the Web is particularly appealing to the small firm. The costs of physical advertising to gain exposure to such an immense group are beyond what all small and many medium-size firms can bear. Further, users of the Web are a highly educated, well paid, computer-literate group, which is an attractive market for vendors of specialty, high-tech products. As usage expands, however, the mean income level of users is slowly going down. Long term, there is little reason to think that the typical Web user will be different from the typical consumer who owns a telephone, television, and answering machine.

Global Presence

The call for companies to become more international has persisted for years, and the Net aids this. Existing multinational corporations such as General Electric and Toyota can provide a uniform set of information and a single image that is available globally. At the other end of the spectrum, companies that serve small niche markets with highly specialized products now have low-cost access to potential consumers around the globe. Without the Web, the costs of locating and selling to enough customers to make business profitable would hamper the probability of these niche companies' long-run survival. Now, a critical mass of a customer base may be accessible.

Supporting Logistics for Physical Delivery

The lure of the Web should not be considered in isolation from other advances in direct retailing, as discussed in the Introduction. Although many services can be delivered on the Web, most products must still be shipped. There has been a quiet revolution in distribution and logistics that allows companies like Federal Express and UPS to efficiently ship small quantities. Airborne Express will even manage a company's inventory, so that shipping a product may be as simple as informing your logistics company (electronically) where to send a particular item. On the East Coast of the United States, for example, some companies can now guarantee next-day delivery for items ordered before 8 P.M. in the evening.

Regulatory Issues

Many companies need to avoid the regulation of their product and service that occurs in the physical world. The most obvious example here is pornography: In the United States, local laws can ban distribution of material that is federally legal, or limit the way in which it is made available. A less obvious example is software, where both export and import laws, or differences in accepted copyright and license agreements, can be a minefield for the small software developer. Controlling distribution via downloads, which can bypass physical controls, may (perhaps inadvertently) avoid regulations. There are also some perfectly legal products that are difficult to distribute because of attitudes and the association they have with morality. A good example is condoms; whereas sales of condoms have boomed due to the AIDS epidemic, selling condoms is still not viewed as "correct" by many retail outlets. It is no surprise that catalogs of condoms, sex aids, and so on have become one of the more successful niche markets in Web-based retail.

Strategic Drivers

What strategic concerns are driving companies onto the Web? Five key drivers, or *push characteristics*, are influencing companies because they emanate from conditions of competitive intensity and a sense of not wanting to be left out. They are changes that no company can ignore.

Drive toward Cost Reduction and Competitiveness through Direct Distribution

Many companies have entered on-line marketing to leapfrog the channel of distribution. Many barriers to entry presently exist for firms attempting to get their new products on store shelves. In the United States, slotting fees have arisen as a new revenue source among retailers over the past 10 or 15 years. Because of the intense competition for shelf space in retail outlets, manufacturers must pay monthly allowances that vary depending on the amount and location of the shelves that their product will occupy. Products shelved at eye level or on end-aisle displays must pay prime slotting fees. Those that occupy the bottom shelf in a grocery store, where buyers do not tend to glance, pay lower fees. Similarly, other forms of push money are required to ensure that one's product gets attention from retail clerks or is not stored in the back of the retail location.

The Web market is (at present) relatively unbiased—anyone with Net access and a share of a Web server can set up shop. Slotting fees and push money are thus not required. Companies can offer their products and services at a cheaper rate to consumers by avoiding the problems of dealing with distributors. Costs of communication, training, and positioning the product to the distribution channel are a considerable addition to overall costs, and are frequently passed on to consumers. These cost reductions allow Web retailers to increase their margins although some new costs are incurred from the investment in computer hardware and software, network access, and Web skills. At present, however, these expenses are minimal compared with conventional distribution and communication costs.

Reducing Reliance on Channel Partners

Aside from the issue of cost, a real threat to a firm's strategic positioning occurs when it must rely on intermediaries to represent, promote, and handle its products in the marketplace. Necessity requires this for small companies that cannot afford to send salespeople to every physical retail store, or conduct store audits to ensure that the product is properly merchandised. Now, however, the Web venue allows the manufacturer to retain control over products until they reach the purchaser.

Many small companies see present distribution channels as crowded, and thus access to the market is limited. Shop shelves are already crowded with existing products. As an example, sales of salsa in the United States now exceed sales of ketchup, but you would be unlikely to deduce this from a quick look at the shelves of grocery stores.

Efficient Use of Advertising Dollars

Given the heavy advertising costs that many small companies incur reaching local markets, on-line marketing is a dream. For as little as $20 a month, a company can have its Web pages hosted by an ISP. Advertising a new virtual shop may simply involve getting listed in Yahoo!, Lycos, and other on-line directories. Compare this cost to approximately $1,000 for a half-page print advertisement in a local newspaper, or as much as $50,000 for a similar advertisement in a nationally recognized paper.

Direct Access: Responsiveness and Learning about Markets

The Web takes direct marketing one stage further—companies can have complete control of their catalog and marketing without needing to rely on printers or postal services. Further, in a technology where they can observe the visitors to their catalog, on-line marketers can track which products are gaining the attention of most shoppers, which ones are being shopped but not bought, which ones are not comprehended, and which ones are purchased outright. This allows retailers to vary their product mix, and to change and update catalog text to better match customer needs and inquiries, in a real-time environment. Physical catalogers, on the other hand, make changes to their catalogs in a batch mode. And, while they have become expert at understanding what buyers buy, they cannot track what buyers look at and question but do not buy, or what buyers bypass immediately in the catalog's pages.

Physical Distribution Opportunities

Where products and services can be distributed over the Web, the reduction in distribution costs allows companies to lower transaction costs (possibly passing some savings onto the consumer) and provide better service. The obvious example here is software retail, which is rapidly moving toward Net-based distribution. The vendor does not have to produce and mail the disks containing the software, and the consumer does not have to struggle with decompression and physical installation of the code. Netscape has in fact changed the game of software distribution, probably forever. You can always download the latest *beta versions* of its products and use them. If you like one of them, you can pay for it when Netscape upgrades to a full-release version, or return to using the old version. The time to market is greatly reduced.

Other products that can be distributed over the Net include company reports, electronic art, and photographic images. The distinction between information-based and physical products is becoming blurred as physical goods such as photographs can exist in digital form, and many products (e.g., automobiles) are searched for and purchased based in part on information.

Inhibitors

That's the good news. Companies, attracted by some features, driven by others, have rushed to have a Web presence. Few of them, however, have noted the inhibitors that deter companies, or will in the future compel them to withdraw, or the biases that make it difficult for some companies to compete. No marketplace is free from bias; none exists with consumers working from perfect information. The days of the Web as a market free-for-all are probably behind us.

Limited Bandwidth

Bandwidth is a present and future problem with the Internet, which has been exacerbated by the Web. If you have ever been able to go to dinner and a movie while you wait for images to download to your browser, you have experienced the limits of bandwidth. The number of bits being moved around that are Web traffic overtook E-mail sometime in early 1995. It's as if the highways are suddenly full of truck convoys that far outnumber the regular automobiles. In the same way that software developers find new ways to use up the growing amount of disc space available on personal computers, Web site developers find new ways to use up bandwidth. The time to download multimedia, particularly video and sound, can be considerable for those sitting at the end of a phone line.

This is a consumer demand problem that users talk about at length because understandably they do not want to wait for pages to download; but it is equally a supply problem for businesses trying to locate themselves on the Web. Companies that pay for large bandwidth connections, or locate with an ISP that can provide this, are the equivalent of "just off the highway." Companies that locate with low-bandwidth connections or unreliable service providers find themselves "up a back road." Just when you've thought you'd freed yourself from channel partners, you now have a new electronic one—your ISP. If your ISP doesn't have redundant systems, so that a second one can kick in if a computer goes down, doesn't back up your work, and doesn't have an arrangement that guarantees the quality of its access to the rest of the Net, then you're risking your business.

In the same way that companies have relocated or set up divisions globally to take advantage of physical sea or air routes, companies will locate to take advantage of telecommunications infrastructure. A company physically located in a highly regulated telecommunications environment (say, France) wishing to retail to the rest of Europe may find it expedient to locate its server in the United States, where communications links to Europe are often better than between European countries. Small companies are moving to take advantage of this—some Web-based startups in the northeastern United States have located their pages on servers in the Midwest where telecommunications are less expensive and access times are reasonable from all parts of the United States.

Biases in the Web as a Marketplace

There is no doubt that companies will have to contend with creeping biases in the Web as a marketplace. On-line services such as America Online (AOL) provide their own marketplaces, more easily accessible to AOL users than the Web. Users of Netscape now find links to certain directories and services already part of their browser. And in trials with Web access through cable, Time Warner has the Web browser default to its PathFinder site.

Phone companies like AT&T and British Telecom are now providing Net access. If they develop a large group of subscribers, they can then divert these users to markets other than the Web. This is particularly true with the BT effort, which resembles an on-line service more than simple Net access since users have to implement and use BT-specific software. Local phone companies, that in the United States are still considering their reaction to the growing use of the Net (Baby Bells are presently kept out by regulations covering long distance traffic), could provide low-cost Net connections with added services such as electronic yellow pages. These may be more attractive to consumers than a global entity like the Web.

In the future, just setting up pages on the Web will not be enough to attract customers. A company may have to buy into multiple on-line services, varying its offerings to suit the different services, and incur the associated costs.

Technical Advances as Barriers to Entry

The original language standard for creating Web pages, HTML 2.0, is relatively easy to use by computer literates, and is simple enough for quick conversion from word processors such as Word and Word Perfect. The Web, however, is rapidly moving toward the much more complex HTML 3.0 and the programming language Java as de facto tools. As time progresses, more companies will become increasingly sophisticated in their technical prowess at formulating interesting Web sites. Thus, a small company that had the skills to develop a Web site in 1995 now finds itself lacking the new skills, and the cost of hiring such skills becomes a barrier. At $200 per hour for a Java programmer, development costs can be significant.

Market Segmentation

It is already apparent that the size of the Internet user community is illusionary. Small companies trying to pitch their product or service to an audience of over 30 million have found, in some case, angry reactions. Perhaps the best known example is the "green card" incident in 1994 involving a pair of Arizona lawyers, Canter and Siegel, who posted a message about their services to over 6,000 Usenet newsgroups. The resulting E-mails and mail bombs brought the servers at Canter and Siegel's ISP to a grinding halt. The size of the market does not relinquish a company from understanding their targeted market segments and identifying their potential customers. Canter and Siegel were slammed for not understanding the Internet "community"; in reality, they just couldn't be bothered to learn anything about marketing and markets.

THE WEB FORCES INNOVATION

Like our marketplace paradigm, we can identify certain attractors and drivers that attract companies *to the technology*, or drive them to implement it. We can even identify some inhibitors.

Attractors

Maybe 10,000 journalists can't be wrong—there's got to be something intrinsically attractive about the technology. What is it?

Multimedia Applications
The ability to develop multimedia applications using images, sound, and video has been around for a while. But it's no good developing things without being able to distribute them—books are not much use without libraries and bookstores. The Web has added distribution to multimedia, meaning that anyone can download a video off any server to which the person has access. (Of course, it might take forever, and the quality may be poor, but it's a quantum leap over having someone mail you the videocassette.)

Real Audio—a format for transmitting sound over the Internet—has meant that sound, such as radio programs, can be downloaded and played on demand. Because humans are naturally multimedia entities (we look, see, and hear), the appeal is tremendous.

Any Place/Any Time/Any Way

Real Audio is used in the United States to distribute some of the radio programs created by National Public Radio (NPR), the not-for-profit no-advertising radio system. One of these, called *All Things Considered,* is a favorite of ours and many others. It is broadcast at 5 P.M. EST each day; but it is also converted to Real Audio format and put on a server at this time. Hence, if you miss the broadcast, you can download it and listen to it on your computer. Or if you're somewhere that doesn't have an NPR affiliate radio station, you can hear it anyway. Moreover, the version posted on the server is separated into segments, allowing you to pick and choose which stories you wish to hear. The Net and Web thus have the attraction of "any time/any place"—you get to pull the information to you when you need it. But this is becoming augmented by "any way." You can listen to a program, read a transcript, or see a summary on a news feed. Conversations in newsgroups are summarized in Web pages. Or you can just post a "please tell me about X" on a newsgroup or in an on-line chat room and read what comes your way. Newspapers such as *The Times* of London and the *Wall Street Journal* provide facilities for on-line readers to customize their editions. The amazing PointCast system, which replaces your screen saver with a news feed that you can customize, puts the ability to edit the news in the hands of the Web user.

Pull/Push Information Balance

To a large extent, the attraction of technologies such as PointCast and Real Audio is their ability to help out with the push/pull dilemma that all organizations face. Individuals want to receive information just in time, with the right level of richness to make their decision.

In many corporations, deciding what information to push to decision makers and what to leave for them to pull is a difficult balancing act. Frequently, they receive a massive printed document containing all possible data, but find themselves calling someone up when they want to know a specific item of data. One division of Xerox used to distribute over 12,000 pages of data to over 1,700 people at least twice a month. Beneficial for paper manufacturers and a good workout for Xerox copiers, but not necessarily the best way to distribute the information.

The Web provides a technology for balancing push and pull. Thousands of pages can be electronically published, but can be arranged so that they can be searched. Search facilities also allow users to do searches over multiple Web pages and sites, looking for where information intersects. Updates can be made immediately (rather than printed copies made twice a month or once a year or whatever), and users can be notified of changes. It's no wonder that development of intranets has taken the corporate world by storm. Add intelligent agents into the mix who can identify trends, and sudden changes in data, and the Web becomes a way of delivering the right information to the right person in the right place.

Strategic Drivers

Electronic Publishing and Cost Reductions

The cost of maintaining, printing, and distributing documentation was recognized as a major cost in many organizations well before the Web. Solutions have ranged from using knowledge-based systems to generate specific documents from generic master documents (e.g., law contracts) to preparing CD-ROMs that field engineers use in place of bulky paper documentation. Electronic publishing on the Web is another piece of the cost reduction puzzle, particularly useful for distributing documentation in the consumer environment. Fidelity Investments is an example of a company that has claimed savings in the millions of dollars by making documentation (in this case investment product information) available electronically on the Web rather than mailing printed versions.

Process Redesign

Over the past 10 years or so, many companies have revisited their organizational processes, and re-engineered them to reduce cycle time, error rates, or other key indicators. Advanced information technology (IT) has come to be viewed as an enabler or process redesign, with technologies such as groupware, EDI, optical storage and retrieval, and knowledge-based systems playing key roles in many redesign efforts.

The Web is a powerful enabler. The ability to search for, retrieve, and forward information can be used as the basis for electronic supported processes. Intranets can be used to empower (we don't really like this word, but sometimes it's useful) customer support personnel and others previously provided with limited and overstandardized information. Data General, for example, provides its telemarketing personnel with intranet access to deep product information and notices of future releases, product changes, and so on. In the first few months of operation, one person sold over $300,000 of equipment over the phone following a discussion based on intranet-delivered information. (An obvious question is why not make the product information available outside the company? In some instances, letting the salesperson access and explain the information may be a better process.)

As an enabler of redesign, however, Web technology exhibits one major limitation. Unlike Lotus Notes, there are no facilities for work-flow management, such as automatic routing of E-mails and documents, or generation of reminders when E-mails are not acted on. These can be added via programming efforts, and planned Web-based groupware will provide many Notes-like facilities, but the Web still may not be the first-choice technology where work-flow management is key.

Inhibitors

There are some very real inhibitors to successful use of the Web as a global IT. While these will vary between organizations and, as always, be related to putting too little or too much money into the technology, or hiring the wrong consultants and the like, three key inhibitors face every organization.

Web and Information Overload

Suddenly, your corporation's intranet tells you everything you never wanted to know. You find that a janitor in your West Coast operation has the same last name as you, and you spend 10 minutes looking through his home page. A search on "sales" and "presentation" tells you that 38 different sales presentations have been posted to the Web in the past month, and the smallest is a 112K PowerPoint file. And this is before you point your browser at the rest of the world.

All of us who have spent more than a few hours surfing or searching realize that the gains from the effort can sometimes be negligible. We become overloaded with information, but too often it's not the information we actually want or need. After years in the Dark Ages, where getting at a specific sales figure took two months and 100 lines of COBOL programming, we've arrived in the Renaissance. It is no wonder that many managers would favor a return to the Dark Ages—adding E-mail and Web usage to already overcrowded schedules has not been a pleasant experience for all. For example, we as ordinary Netizens now receive over 100 E-mails a day. These are new communications, not E-mails instead of phone calls or faxes.

Skills and Appropriate Use

While to many of us, the phrase "It's on the Web" signals that we can get at it fast, this is not true for all workers. For many clerical workers, accustomed to detailed training in system use and stringent operating procedures, just dealing with the browser interface can be problematic. Some attempts to use intranets to support internal operations have floundered because of resistance to the technology. Putting the phone book on-line is wonderful (it gets updated more frequently and can be searched), but if the CEO's secretary can't fire up a browser and still uses the hard copy, then it is expedient to keep updating that, too.

Which brings up the issue of appropriate use. The specter of young executives spending valuable company time surfing the Web for entertainment has occurred to management. In companies where the free games in Windows 3.1 were not distributed so that employees could not be tempted to waste their time playing solitaire and minesweeper, Web access is considered to be in the same category as permitting a 16-year-old to be alone in a bar. Our experience suggests that many companies face a major dilemma here: "Playing" is necessary to develop skills and awareness, but inappropriate use can drain time, particularly at the clerical level. As always, any strategic effort to take advantage of the Web must consider training.

Early Adoption

In the General Electric case in Chapter 1, GE faced a number of technical problems and other concerns due to their early adoption of the Web. This is always the way— the leaders pay a higher price, with hopes of gaining the larger benefit. Business use of the Web is currently waiting for many "when" questions to be resolved: When will consumers be as happy to give credit card numbers over the Net as they are over the phone? When will security concerns disappear (if ever)? When will flexible EDI-like interorganizational use be acceptable to most companies? And when will this graphic finish downloading?

Many companies are right to sense new business opportunities, but also right to take things slowly. With telephone and ATM banking now accounting for over 50 percent of all consumer banking transactions in the United States, Web- and Net-based banking is a natural extension. But early entrants in home PC-based banking such as Chemical Bank did not fare well.[3] It remains to be seen whether early movers into Web-based banking services such as First Union and Wells Fargo gain advantage from this relative to followers.

THE WEB AS A GLOBAL AND CULTURAL PHENOMENON

In 1994, General Electric was seriously concerned about alienating the Net community with its use of the Web. Two years later, the number of companies with Web sites exceeded 37,000[4] in the United States alone. Net culture and netiquette concerns are now less prevalent. Instead, for businesses entering the Web arena, electronic commerce, internal and external security, positive public relations, and accurate corporate communications have become points of intense focus.

Those who suggested that commercialization would just wipe out Net culture, replacing it by some generic business-speak, didn't get it quite right. Everything from smiley faces to cyber-speak has been assimilated into our culture with a dizzying speed. We now see advertisers using prefixes like "cyber" in regular print advertising just to suggest the novel features of the product or service (Cybercereal! Virtual flavor!). *Time* and *Newsweek* writers sometimes sound as if they've been reading science fiction since the early 1980s, and even the *Wall Street Journal* has not been immune.

Whether you consider Net culture to be bunk, a burden, or something that has to be understood, there are three key aspects to communication on the Net that any business must consider: *politeness, lack of intrusiveness,* and *emergent groups.*

Politeness

The old adage, if you can't say anything nice, don't say anything at all, is now truer than ever. There is a gap between what you write when you communicate electronically and what others read and understand. Without the richness of human contact and visual signals, misinterpretation is more frequent (which explains the need for smiley faces to convey irony, irreverence, etc.). Many users of E-mail do not understand that:

1. Messages can be forwarded to others. Even if you use blind copy or say "don't forward" in your E-mail, there is nothing you can do to stop the receiver from forwarding

your comments. Don't write anything that you're not willing to repeat in public.

2. Many companies back up all incoming and outgoing E-mails. Ostensibly done to secure people from server failure, this has the added advantage (or disadvantage) of making old E-mail communications accessible to the likes of lawyers.

One secretary sent her colleagues E-mails that lampooned her boss. She used a group name that had been set up for messages to all clerical staff, unaware that anything sent to this group name was also copied to the boss.

Lack of Intrusiveness

Lack of intrusiveness means that people don't send uninvited E-mails to people they don't know, do not intrude into Usenet and on-line service discussions with unsolicited comments, and generally behave themselves. If you're a right-wing Republican, you don't blunder into left-wing Democratic political discussions and just slam everybody. Virtual space is treasured. (You're free to monitor all publicly available discussions, which as discussed in Chapter 5 can be of great value.) Unfortunately, some instances of intrusiveness are certainly unethical and may even be illegal, but are rarely detected. We ran across one small company (a producer of Unix software) that maintains a second domain name (that sounds nothing like its company name) and an associated set of E-mail accounts just for taking part in Usenet discussions. When someone posts a message asking for software suggestions, this company posts testimonials from "customers" saying how "excellent" the company and their software is.

Emergent Groups

Perhaps the most remarkable aspect of Net culture is that of *emergent groups*. These groups of individuals emerge, over time, to become contributors to either a Usenet newsgroup, an on-line service discussion group, a chat room, or a list server. They share a defined interest or issue; these are the people who have made the *Star Trek* Usenet groups some of the most heavily used newsgroups.

Unlike groups or teams that may be formed through some formal mechanism, and then disbanded, emergent groups are dynamic to the extent that even the number of members at any one time may be impossible to define. Even working out what the key discussion items are can be difficult. (We have a family member who subscribes to AOL's needlework discussion group, where the discussion is dominated by women's issues far beyond needlework.) Some may be key players who drive discussions, others just "lurkers" who read but never contribute.

Companies with intranets or Lotus Notes are just beginning to discover the power of emergent groups; workers migrate to the group where they can make the biggest impact. Some Web marketers now talk glowingly of creating a "customer community" where customers can share experiences or idle chatter via chat rooms, with customers returning to an on-line service or Web page to be part of this community. This has always been an aspect of Net culture; since the advent of the Web, many arrivals have taken a long time to work this out.

In many ways, on-line services are a better forum for consumer discussions and emergent groups than are the Usenet newsgroups. Access to Usenet is not as easy as access to the discussion groups in on-line services, which may be available through simple menu options (compare this with having to use the news reader in, say,

Netscape Navigator). These services present two problems for business:

1. The on-line service has complete control over content. When AOL decided to disallow all postings with the word "breast," members of a discussion group on breast cancer were rightly outraged. CompuServe has been involved in litigation in the California courts; it was sued by an individual for failure to curb flaming about his company and products.

2. The services are able to collect all discussion material, perhaps to analyze, redistribute, and sell it in the future. Like airline reservation systems in the early 1980s (prior to being forced to share information across airlines), they may end up with control over the information that can be generated within their systems.

Global Brands, Image, and All That Stuff

To a company that does business globally, managing a Web site is both an opportunity and a problem. If a company has a global brand name that it wants to instill throughout the world, having a single Web site can further this goal. Examples include British Airways and Federal Express. Most companies, however, have different operating policies, slightly different products or services, and even different divisions across geographic regions and countries. Thus, even a company as universal as McDonald's serves grits and biscuits in the southern United States and varies its menu from country to country. How do you deal with these differences when on the Web? Federal Express has had to address this issue since its service varies differently between the United States and Canada, let alone the rest of the world. And it isn't easy. IBM decided to maintain different Web pages for its different regions, and visitors to the main IBM site are asked to identify the country they're in. The server then redirects the request. A number of Japanese companies, including Canon and Hitachi, have developed different sites for their home market and their biggest export market, the United States.

In 1995, the British company Virgin Atlantic Airways probably became the first company to be fined for information on its Web site when the Federal Aviation Authority (FAA) objected to the way it put prices on its site. And there's the crux—*pricing*. If there's one thing that differs across countries, it's the way items are priced and offered for sale, including the regulations covering offering, contract, and complaint. As in the GE Plastics case, many companies have decided to keep prices off Web sites. Present a global image, reinforce your brand name, but don't open up the hornet's nest. It may be global marketing, but as yet it isn't global commerce.

Local Culture, Local Infrastructure

On the Web, global branding meets local culture head on. Visitors to any Web page are viewing the page under the weight or enlightenment determined by their education and social conditions, using their own country's technological infrastructure.

The most obvious manifestation of culture is language. The lingua franca of the Net and Web is English, but many companies outside English-speaking nations have chosen to present Web pages in their national languages. If you're in France and your target market is only in France, then your pages will be in French. Some non-American global companies have obviously struggled with this: Siemens's main Web page is in English, but many of the subpages are in German. A number of Asian companies, of which Ricoh in Japan and Samsung in Korea are perhaps the best examples, manage pages in both their native language and English.

Culture, however, is a lot more than language. It is an untested but widely accepted hypothesis that Web pages reflect differences in local attitudes and culture. Thus American Web sites are highly visual and "in your face," British and Australian are whimsical and humorous, German sites get quickly down to business with no peripheral detail, and Asian sites are polite and unobtrusive. For the Web site developer, the extent to which a site will be visited by overseas visitors is an important consideration that has eluded many. Making something attractive to U.S. teenagers may make it seem outlandish to middle-aged Germans. This may not be intrinsically bad, but global presence should not be predicated on local norms.

In a global market, even simple data management tasks are important. Consider the simple global concept of day, month, and year. In the United States, dates are usually written month/day/year, such as 8/23/96. In Europe, day/month/year is more common, as in 23/8/96. How many Web pages exist that blindly write dates in numeric order, assuming the cultural interpretation that exists locally is universal? One way to spot that some thought has gone into design is to see how dates are handled: August 23, 1996, or 23 August 1996 (i.e., text for the month), should always be used.

A frequently discussed aspect of local national cultures, at least in the United States, has been infrastructure. The United States has the most advanced telecommunications infrastructure in the world, and is thus at the forefront of development. Coupled with aggressive pricing at all levels, which is aimed at generating market share before profit, the infrastructure is not just advanced, it offers fixed prices and is affordable. Companies in Australia are limited by variable pricing and taxes on telecommunications. Those in Europe and Asia are held back by regulation and lack of competition, and many other countries don't even have a reliable phone system.

While there are no doubt differences among various countries in National Information Infrastructure (NII), the United States has the most advanced and (possibly) least regulated phone system in the world. Other countries have developed local systems that have delivered information to households and businesses for years. In the United Kingdom, the Teletext system provides pages of text information that are broadcast on the back of television channels. A user must have a Teletext TV (more expensive than a regular TV), but has no other costs after purchase. Up-to-the minute sports scores, travel news, and stock quotes are available (quite similar to the offerings of many Web-based news efforts). There are no graphics, and interaction is limited to the buttons a viewer can push on a remote, but, heck, it's nearly free.

Some Teletext pages are, to a Web literate, reminiscent of virtual brochures on the Web. (See Chapter 7 for a fuller discussion of virtual brochures.) Comet, a large U.K. electronics and household goods retailer, maintains a Teletext page that lists current prices, special offers, and news about their stores. What's more, Comet puts its Teletext address on its print advertising. Sounds familiar? The penetration of Teletext in the United Kingdom is greater than that of the Internet (although the number of Net users will probably overtake the number of households with Teletext soon). Hence a company expecting to advertise or sell through the Web as a sole channel will have to contend with competition from a medium it is not even in.

REVISITING THE VALUE CHAIN

Our argument is that the Web can be viewed as a market-place, an enabler for global IS, a cultural challenge, or (more likely) some combination of the three. Others have used various other analogies or methods to help contribute to understanding, particularly value chain analysis. Here, we'll have our say on this matter.

Depending on what you've read, the impact of the Net on the value chain is either support for many stages of the chain, the collapse of value chains, or the leapfrogging of downstream operations by business (as in direct sales).

Our view is that the Net has two effects on value chains. First, as when Nine Lives provided agents for shoppers and when GE Plastics put its technical specifications on the Web, it opens up new ways to add value to existing services and products. Whether this substantially changes a value chain or not depends on the importance of the additional value-added service. For some companies, the added value may be their differentiation in the marketplace, and thus the service itself becomes a major part of the value chain. Start-up Web companies selling CDs over the Web find themselves in this position. The CD itself is really a commodity; the buyer will be influenced by pricing, but also by the search or recommendation facilities of the Web site (perhaps what might be called the virtual shopping experience). The value chain is largely virtual.

Second (and many have commented on this, particularly Rayport and Sviolka at Harvard[5]), the Net is leading to a desegregation of content, context (essentially the methods whereby the content is delivered), and infrastructure. This split is best understood by an example, and at present publishing is the most advanced example.

Old black-and-white movies about journalism always seem to have a scene where the reporter runs into the press-room and shouts, "Stop the press!" You then see the presses roll (presumably after they've been stopped and the type reset), newspapers coming off the line, delivery trucks throwing bundled newspapers onto the curb, and then newspaper sellers shouting, "Read all about it," or some such thing. Finally, we get to see the headline, usually after a shot where the paper spins around really fast before coming into focus.

That was in the old days. Today, many papers and magazines have a Web site. Versions of *Time, Newsweek,* and *Sports Illustrated* are available through on-line services. The content is delivered through multiple mediums. And the supplier, such as America Online, does not necessarily own the infrastructure that is the means of delivering the content in the context of the service. News sources, on-line services, and Web sites, plus networking, combine in complex ways to deliver the content. The result is a focus on one part of the disaggregated value chain: *Newsweek* put the content together, AOL arranges for consumers to view it, and companies like PSI and Netcom (plus your local phone company) make the networks run.

This separation of content, context, and infrastructure existed prior to the Web, but is more blatant in the virtual world because the linkages between these aspects are invisible. When we focus on content, sometimes context and infrastructure can be taken for granted. Hence anyone can be a publisher on the Web—we put together our magazine, announce it on a newsgroup, and just sit back while the context (access through the Web) and the infrastructure does its stuff. In the physical world, we can get a magazine printed and distributed for us, but this demands considerable coordination between our content, the context (e.g., a free news sheet, or a glossy magazine), and infrastructure. Coordination will have to take place over time, financially, and probably contractually. Admittedly, other business areas are not as obviously disaggregated, but the point is this: In the virtual world, coordination between content,

Content	Editorial, Videos, Transaction services, Payments, Bookings, Notification
Context	Web sites, On-line services, Software agents, Mailing lists, Discussion groups, Intranet and other limited access sites.
Infrastructure	Networks, Routers, Local phone systems, Servers, Browsers, Additional plug-ins

Figure 2-2. Disaggregation on the Web.

context, and infrastructure is more straightforward. We pay for access to the infrastructure, but not at a rate dependent on what we use it for; we may be able to control our context purely through choice of service provider; we are left to focus on the content. Thus, we may spend less energy in coordinating the components.

The Web and the Net are in some respects an experimental laboratory for this desegregation (Figure 2–2). We combine content, including transactions and payment, and then deliver it through a context that combines Web sites, discussion groups, and access procedures. The same content may be distributed in numerous ways, for example, in an electronic newsletter, on a Web site, or through an online service.

THE WEB AS A STRATEGIC ADVANTAGE: DEFINING OBJECTIVES

The big question is: How do we turn Internet and intranet-working into business advantage? Table 2–1 lists 11 ideas that combine many of the concepts that have been discussed. They are key *objectives* for using the technology

Table 2–1. Opportunities for use of the Web.

Objective	Summary
Centralize and distribute	Put items (such as documents) into digital form and distribute internally and/or externally.
Reduce cycle-time and control the message	Reduce the time to publish, discuss, or offer new services. Control distribution yourself.
Build an information supply chain	Deliver the right information in the right place to the right people. Move away from generic position.
Benefit from emergent teams	Let teams emerge rather than be created. Think of teams as dynamic and self-driven.
Open a virtual store	Advertise, market, and deliver through the Web. Go global.
Support customer decision making	Gives customers the tools to make decisions. Don't just communicate; empower them.
Create electronic inter-organizational processes	Create systems and processes to interact with customers, suppliers, and competitors. Provide added value. Move beyond EDI.
Build shared inter-organizational systems	Build systems that share knowledge with partners and customers. Become the airline reservation system for your industry.
Build competitive inter-organizational systems	Build systems that make suppliers or competitors compete. Make bidding, certification, etc., electronic.
Develop an electronic market	Provide an electronic service based upon brokerage or market principles.
Collaborate to create new businesses	Collaborate with new partners in virtual enterprises. Look for new ways to combined services.

that can be driven by business strategy or that lead to new business opportunities. This approach will help you focus on a technology opportunity without being limited by specific business requirements or functions.

The objectives summarized in Table 2–1 range from ideas that can be quickly acted on (with specific resources and identifiable results) to ideas that require a major commitment of funds and, more importantly, determination.

Centralize and Distribute

Use the Web to centralize and distribute items in electronic form—documents, software, images, sound. This can range from software fixes distributed to customers to internal newsletters distributed over an intranet. Universities provide on-line catalogs, Sun Microsystems provides software fixes and updates, NPR produces its program *All Things Considered*—these are all examples of this common objective.

The result is a reduction in distribution costs plus an ability to deliver items quicker, globally, on demand, and (sometimes) in a more usable format. The key is managing the central repository. Database technology is vital, and processes have to be in place to make sure that updating occurs as and when necessary.

Reduce Cycle Time and Control the Message

Aim for massive reduction in the time taken to deliver items while controlling the message. This can range from announcing and providing information on a new financial product over the Web, to contacting and distributing information to customers (perhaps bypassing, for example, print journalism, journalists, and industry insiders). Financial institutions like Fidelity are using the Web to offer new mutual funds (unit trusts); IBM is using it to make product announcements; Adobe puts its product strategy on-line as it announces it to analysts.

The result is faster time to market, more direct interaction with customers, and a better ability to control the way in which the company gets its message (and subsequent products and services) across. Interaction is vital here (particularly E-mail communication with customers), and the message must be consistent across all media, both virtual and physical.

Build an Information Supply Chain

Use the Web to deliver timely pertinent information to individuals and teams. Think in terms of information supply, not just physical supply chains. Intranets are still mainly pull information sources; use agents and tools such as Real Audio and PointCast to deliver specific customizable information gathered from internal and external sources. Everyone in an organization can have the facilities of an Executive Information System (EIS) at the fraction of the expense of proprietary EIS. Wal-Mart is using agents to alert mangers to sudden changes in sales; GE puts supplier contracts on the Web, allowing managers to search for supplier contracts across all divisions prior to entering into any new agreement.

The result is more efficient working practices based on the right information. Providing the internal infrastructure and managing the ability to personalize the information supply are both vital.

Benefit from Emergent Teams

Stop creating committees and arbitrary teams. Let teams emerge through collaboration around a discussion group or Notes database; let leaders emerge from those who lead the discussion. Albany International, the global manufacturer of paper machine products, used Notes to let emergent teams determine the company's PC and software requirements. Discussions were terminated only when a broad consensus was reached.

The result is the ability to benefit from a fuller range of personnel involvement, making the best use of individuals. You need to control the process and make sure that emergent teams do not become the new committees. Easy access to virtual meeting places—be it Usenet discussion groups or Notes databases—is essential.

Open a Virtual Store
Market and retail your product or service through the Web. Set up a Web site that uses virtual retailing paradigms (see Chapter 7) to sell. Use this as a method to move inexpensively into direct sales, or to supplement existing direct sales channels. Use the information gained from the Web server to better understand and model your customers. CD Now and Amazon.com are building businesses based on immediate availability and ordering of, respectively, any CD or book. Niche products from wine to hot sauce can be sold to a wider customer base. The seed catalog company Burpee also sells through a Web site, allowing it to provide special offers on slow moving products. When Harley-Davidson of Stamford, Connecticut (a dealership), opened a Web site, orders come in from customers as far away as Russia and Singapore.

The result can be an increase in sales, and, particularly for small companies, an extension of geographic reach. Integration with transaction processing and logistics is essential and the design of appealing Web sites that attract customers and keep them returning is important. The downside is that a company must be ready for intense competition—in a global marketplace, another company in another country may have a price or culture advantage.

Support Customer Decision Making
Don't just sell to consumers—support their decision-making process. Help them work out which flight to purchase a ticket for, which vacation to go on, which mutual

fund to purchase. L'eggs provides a site that will estimate panty-hose size given other information. In the United Kingdom, the Abbey-National bank will provide a detailed mortgage estimate for all types of mortgages, based on personal financial information. In the GE Plastics case in Chapter 1, the technical specifications provided through the Web site are ultimately used to support the decision "which plastic?" The Belgian chocolate maker Godiva allows visitors to its Web site to set up lists of birthdays; an agent then E-mails reminders to the visitor.

If you can provide value-added services, particularly when selling commodity products or operating in a highly competitive market, then you increase your chance of retaining customers. Understanding customers and providing them with a valuable service is vital. Customer decision support can be considered a mature stage of Web use that few companies have yet achieved.

Create Electronic Interorganizational Processes

Customer decision support is really just one type of electronic interorganizational process. Other possibilities include allowing customers to generate complex orders and have them verified immediately, EDI-like supplier interactions, and reconciliation of funds. Think beyond the standards and limitations of EDI; think about building one-to-one interactions with specific suppliers, partners, and customers rather than generic interactions. Many manufacturing companies are working out ways that allow customers to specify orders for customized products, not just order off-the-shelf products. Hewlett Packard has experimented with allowing customers to schedule production, in much the same way a customer for a haircut might schedule an appointment with a stylist.

Companies that move beyond the limitations of EDI will gain an advantage. But providing the security and reliability of existing EDI structures is a major technical challenge.

Extranets, which use Internet technology, but are not publically accessible, go some way toward this.

Build Shared Interorganizational Systems

What do you have to do to become the airline reservation system of your marketplace? What systems can you build that will allow you to provide shared information to customers? In many areas, such as banking and insurance, companies have rushed to set up Web sites and services that reflect a very internal view of the company. Advantage will flow to companies that build *the* site for *their entire industry.* Become the place to go for financial or banking advice; recommend other services or products that are not your own. As with airline reservation systems, profits will flow to those who control the market. Doing this takes a major commitment and a long-term view. Shortsighted companies that will not even put links to competitors on their site ("I'm not sending visitors to a competitor!") probably need a brain transplant before undertaking such endeavors.

Build Competitive Interorganizational Systems

The opposite of building systems that share and distribute information is building ones that increase competition, either through sharing or hiding information. GE uses the Net to increase the number of its suppliers and competition between them. Specification of commodity fabricated parts, including CAD diagrams, are available to certified suppliers through an extranet. Bids are accepted electronically; at present, suppliers can see the lowest three bids. GE is not bound to take the lowest bid. Since EDI has reduced the number of suppliers in many industries, in part because of the expense and inconvenience of EDI to many small firms, companies are now looking for ways to increase their supplier base.

The results of competitive systems can be lower costs. However, the same implicit quality processes that exist in the physical world must be implemented in the virtual world. No manufacturer wants non-quality-assured suppliers bidding.

Develop an Electronic Market

Build an electronic market where customers and providers can meet (Chapter 7 discusses types of electronic markets—especially electronic auctions and brokerages—in detail). Manage the market process, including contracts and payments. Move physically slow, inefficient markets onto the Net.

Aucnet in Japan is an example of replacing a physical market with an electronic counterpart. Used car auctions, where dealers had to gather at a limited number of physical locations to view cars and make bids, was replaced by a nationwide system that allows dealers to see the cars and written details electronically, and then make bids through a teleconferencing system. When the growers that run the Dutch flower auctions acted to keep flowers imported from Africa out of the auctions, the East Africa growers created a virtual auction allowing buyers to see and bid on flowers through PCs linked to a flower warehouse.

Despite few examples, it is easy to imagine how this might replace some physical markets. Consider buying a house. In the United States, many purchasers hire a realtor who then hunts out suitable houses, normally through a local database. Each city typically has a multiple-listing database, and some of these are now on the Web.

In the United Kingdom, prospective purchasers must waste days interacting with many small estate agents (as realtors are called in the United Kingdom). It is not unusual for even a small village to have a local estate agent who sells the houses in that village. In an electronic market, information on all available properties would be available to the purchaser through the Web. Search mechanisms would be employed to find properties; E-mail to gather more details; videoconferencing for virtual visits. In the United Kingdom, a number of estate agents have built sites that simply extend their present marketing (i.e., details on only their vendors). Attempts to build cross-agent sites (such as The Property Wave and similar Web sites) have only attracted a few small agents, thus suffering from limited product inventory. Innovative use of Internetworking would focus on the customer and recognize that an electronic market should reduce the search problem.

Like the airline reservation analogy described earlier, whoever builds a successful electronic market will have a substantial advantage. Due to the existing structure of many industries, such as the housing market in the United Kingdom, "buy in" from many businesses is necessary. This suggests a collaborative effort, or a new entrepreneurial effort that can quickly persuade many existing businesses to take part.

Collaborate to Create New Businesses

In the virtual world, collaboration to create new businesses may be easier than in the physical world. Collocation need not be necessary; desegregation makes for interesting bedfellows and new opportunities. Thus, Microsoft can create a new quality magazine *(Slate)* since it need not purchase or coordinate with a physical distribution system. NBC and Microsoft combine to create MSNBC, merging NBC's content and context systems (news and television programming) with Microsoft's context system (Web site and the MS Network (MSN)).

In the future, expect such collaborations to be commonplace. Music companies will collaborate with virtual stores to offer special or prerelease versions of CDs. Financial newspapers will collaborate with security exchange companies—if you read a review and like the stock, click on the text to activate the purchase. Educational institutions will find themselves having to collaborate with publishers and consulting firms to deliver tailored education to executives. In the earlier example of an electronic market for houses, a nationwide residential cleaning firm might collaborate since people moving from one house to another are a key market for their services.

Perhaps the bottom line is this: Some or all the predictions for business use of the Net and Web could be wrong since we don't know what new markets and business opportunities will emerge. In 1980, the PC market was in its infancy, and Federal Express was still considered a start-up.

THE WEB AS A STRATEGIC NECESSITY

Chapter 3 discusses the efforts of United Technologies Carrier, one of the two largest providers of air conditioning equipment in the world, to build a "resource center" for its industry (in effect, what is referred to in this chapter as a shared interorganizational system). One of the driving forces behind the decision to undertake this effort was the development by Trane, the other major air conditioning company, of a Web site.

For many companies, using the Net and Web to compete is a strategic necessity. Seeing competitors make outrageous claims for their Web site has compelled many a vice president to do likewise. One company in California that we are familiar with would not fund an internal $200,000 Web project in early 1995; by the end of that year, they had committed $1 million to the work.

For many companies in mature markets, however, having a "home" on the Web may be irrelevant. The big change will be the use of interorganizational systems and electronic markets. Companies have to work out how to react to such changes now—in the real estate example given earlier, what should be the position of a small independent real estate company? Of a large company owned by an even larger financial company (such as the Halifax Property Services in the United Kingdom)? Despite the paucity of many present Web efforts, experimentation is vital. A company has to determine if it can compete technically: Are internal skills or are external contracts necessary?

We have suggested to many companies that, in addition to generating some kind of Web presence, they form a team or consulting relationship that looks to the future. Not only should such a team consider the opportunities available within the Web realm, they should consider the threats that may arise from a competitor moving within that world first. If your competitor tries to build an electronic market for your industry, will they kick your butt or lose their shirt? Or more importantly, what can you do to make sure that they lose their shirt? Any team that considers competitive reactions must not be dominated by functional areas. Ask the IS people to do it, and you may get a discourse on the role of IT standards. Ask the marketing contingent, and they will concentrate on advertising. Instead, find a bunch of 20-year-olds who have all played Doom and Myst, and get them to paint a picture in front of the marketing and IS people.

THE WEB AS A STRATEGIC DISADVANTAGE

For some companies, Internetworking will prove to be an ultimate disadvantage. Well-planned efforts will turn into disasters; opportunities will be missed or messed up. Executives will talk about the "good old days" before the Net and Web.

Even the present state of affairs, with many limited Web sites just providing some marketing information, offers opportunities for your Web site being used against you. If information is going to be made universally available to all, an analysis should be done of what harm could occur through access to the Web site. We're not talking security problems here, just the plain results of competitors having better access to your company's information. When we asked a number of executives to give us an example of how the Web was valuable to them, this reply from a sales leader in a company that competes against Silicon Graphics provided much food for thought:

> Being in the highly competitive sales environment, I use the Web to gain a competitive advantage. When I am in a competitive situation with a vendor, I can do a search on the company to read about its products, services, and company direction. Recently, for example, I was competing against Silicon Graphics and did a search to get their product specs to compare against mine. My competitive advantage was that I had up-to-date, real-time information that I printed and read on my way to the sales call. I was able to talk intelligently about the competitor without investing much time at all. My company spends lots of time and money doing this kind of research to provide to the sales force and it is often out of date before we get it. The Web solves this problem!

In the words of a colleague, IT can do things for you, or to you. Or sometimes both.

3

Re-Engineering the Re-Wired Corporation

Remember the early 1990s and all those management gurus who told us that if we would only re-engineer our companies we would all be fabulously wealthy and famous? Remember thinking . . . "Re-engineering? But, we never engineered the corporation in the first place." Then we tried to put those lofty ideas into practice and what happened? Many of us lost our jobs. But the idea of re-engineering—organizing people around the flow of information within an organization—is a good one. In fact, if you read about the new breed of companies and how they are organized, they are indeed engineered from the beginning. And many of them are using Internet technologies to focus their flow of information and provide their people with a great wealth of information to keep them more productive. What we hope to do in this chapter is highlight some of the ways in which the Net can help engineer and re-engineer, provide some metrics and measurements and give some concrete cases where the Net gain outweighed the Net loss (pun intended).

RE-WIRING AND THE THINKING ORGANIZATION

The title of this book suggests that to make your company more efficient you may need radical change, lots of money, and a team of consultants (primarily made up of the authors of books calling for radical change). While that last bit may be true, we want you, the manager, to understand the fundamental difference between re-engineering and re-wiring. Re-engineering is the generic term for re-structuring your business along the flow of information within your business. Re-wiring is the process of structuring your business to connect as many employees as possible to both corporate resources, on an intranet, and to outside resources, on the Internet.

The connection is what is important. The connection changes the nature of your company from one of those stuck-in-the-mud knowledge organizations into a *thinking organization*. Knowledge is stored, static, and archived. Knowledge workers are those people who know just enough to be dangerous. Thought is dynamic, continuous, and alive. The connection bridges the gap between knowledge and thought. Smart people are not those that remember every detail of everything but those that know where to look for knowledge. By the simple act of connecting, you allow those smart people access to your company's vast store of knowledge. Thinking also implies something beyond knowledge—interacting. The connection also makes it possible for your entire workforce to interact; get to know one another, share ideas, work on making the company better.

The Totally Re-Wired Company—Metcalfe's Law

Why, in this era of whatever-sizing, should we be concerned about connections? Just when you thought growth was the economic animal you were stalking, along comes another economist saying that ideas are what fuel the engine of growth. So without ideas, there is no growth. And without growth, there is no profit. And without profit it becomes time to dust off that old resume. How then do we generate ideas and capture them, in effect, becoming a *re-wired corporation*?

Bob Metcalfe invented the *Ethernet*.[1] For you nontechies, the Ethernet is what allows our packets of information to flow out across the network and arrive at the right place, in the right order, at the right time. Mr. Metcalfe turned his idea into 3Com, one of the industry's most successful networking companies to date and a dominant supplier of network interface cards and adapters. It is safe to say that he has played and continues to take a major role in the development of networked computing. When Mr. Metcalfe expounds on his ideas, people generally listen. Perhaps the most significant of these ideas is Metcalfe's Law, which states that for any communications network (this is not just limited to computer networks), the number of nodes (n) on a network yields that number squared in potential value, n^2. So a 2-node network gives 4 times the value. A 10-node network gives 100 times the value (see Figure 3–1).

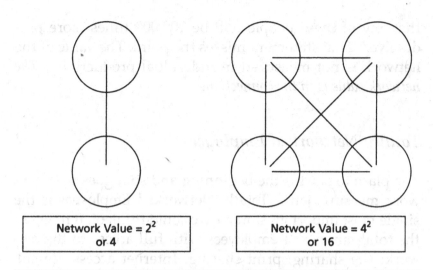

Network Value = 2^2
or 4

Network Value = 4^2
or 16

Figure 3–1. Metcalfe's Law.

This law really takes off when you start playing around with large numbers. So if you have 100 employees with connected workstations, you have a network with 10,000 times the value of a single machine. The value is derived from the possibilities that exist because of the connection. A network allows file sharing, print sharing, E-mail, Web browsing, Usenet newsgroups, and connections to the outside world. But this example includes just computers. Now let us assume that each of those 100 employees also has access to a telephone. There lies another network with at least the same value (100 squared) if you only use the phone network internally. Combine the two networks and you have a potential value of 40,000 or 4 times the value of each network separately. Add to that some network faxing, copying, and Internet access and suddenly your 100 employees have access to a network that can easily be worth 50,000 to 100,000 times the value of their stand-alone machine. But, you say rather cynically, "That does not mean that any of these people will be 100,000 times more productive." That statement misses the point. The value of the network is not measured in individual productivity: *The network value is in the connection.*

Totally Networked Employees

The place to start is the beginning and so it goes with network measurements. Totally Networked Employees is the single most important statistic we could think of. It refers to the total number of employees with full access to the network: file sharing, print sharing, Internet access, E-mail, Newsgroup, Web, intranet, remote access accounts, the complete range of use. The ideal measurement reaches 100 percent. So do not be surprised if you fall somewhere short.

The formula is simply:

- Total network connections
- Divided by
- Total employees plus total employees with remote access
- Times 100 (the percentage thing)

or

$$\frac{\text{Total network connections}}{\text{Total employees} + \text{Total employees with remote access}} \times 100$$

Here's how the formula works:

- *Total network connections* is the total of every PC, printer, copier, water cooler, and Coke machine connected to your internal network or networks (intranet). This gets tricky when you start to figure out remote users, since they generally dial in to a modem pool. The pool may have 10 modems that actually support 50 people. Count the 50 people as the number of connections.

- *Total employees with remote access* have two network connections—their remote machine for travel and their LAN connection for office use. That is why they are special and get counted twice.
- *Total employees* is simple. Just add up everyone who works in your company. And that means everyone, not just the ones you think are really working.

Reading the Results

The ideal measurement is 100 percent. This means that for every person there is a connection, whether they are in the field or in the office. Measurements over 100 percent are also good, unless they reach over 150 percent. This indicates a reserve of bandwidth and the capacity for growth. Anything over 150 percent though is a bit excessive. For example, at One World (the author's design firm) we have a metric of nearly 350 percent, indicating that every person has over three network connections. Silly at best, but what the hell.

Now for the not-so-rosy scenario: companies with measurements under 100 percent. This indicates a potentially unproductive state in which people who may need access to the network are not receiving it. From the re-wiring perspective, this means that some people in the organization are off-line on the information stream. You can expect information to be pooled in this scenario. People who need that information probably have to wait in line, fill out forms, hassle IS people just to get the basic information to do their jobs. These people are frustrated because every time they have a good idea, they cannot get to the information that they need to make their ideas work. To find out if your company is like this, go to the IS people and ask for a report that you know does not exist but that you know the parameters of. See what their reaction is. Our guess is that you will be stonewalled, and stone walls are meant to be broken.

Measuring Return on Investment

We all like to know that we are getting what we pay for and our Internet and intranet presence is no exception to that rule. If you buy into the hype surrounding the Net, you probably won't have any problem shelling out mega-bucks in search of yet another technology fix to business problems. But if you are like most of the people we know, you are approaching these technologies with a grain of salt, a dash of skepticism, and a bucket of fear. New technologies usually mean a sinkhole into which you pour ever increasing amounts of cash, only to find that there is little way of recovering your investment much less measuring any return.

There are two approaches to determining investment in the Net. The easiest way is to consider the network as basic infrastructure and treat it as overhead. This approach, while simple, makes a great deal of sense. We think that providing a basic network connection is invaluable to any corporation.

The second approach is to break the Net up into components and determine costs based on how the technologies are being applied. For example, your communications package now becomes "software" packages that are costed to IS, the same way that any other application such as a word processor or spreadsheet would be costed.

Next, take each specific area addressed by part of your Net and assign it a function. For example, if you have a customer service portion of your site that is designed to answer users' questions, assign it to customer service. Now you have to determine how much that portion of the site gets used. This can be done using a combination of log files (determine the number of hits) and actual requests initiated or sent via the Net. So if someone calls your support center who found your phone number on your Web site, that should be attributed to the cost of the site. If you receive E-mail from a link on your site, that's obviously Web related.

The site itself costs something to develop and maintain. Usually there is some corporate component that points people in the direction of the functional areas, a kind of home page if you will. Then there is the cost of the server and the telecommunications equipment associated with keeping the server connected to the Net. This costs real money, but like a phone system, the usage and costs should be spread out throughout the organization and the groups using the network. Again, it is helpful to assign functional areas to the tasks being performed by the servers and the telecommunications gear. A mail server used by the entire corporation is probably best paid for by a corporate contribution, or it can be broken down on an account basis. Marketing uses 1,000 accounts out of 2,000 in the corporation? They should bear half the cost burden. The same is true for Web servers. Your programming division maintains a download area that accounts for half the disk space? Charge them half the cost. The costs that you determine become each group's contribution to the overhead of the network infrastructure.

From here, the math is simple. Add up your costs for producing and maintaining your portion of the site and add in a contribution to the overhead of the site. This should include any labor that you put forth from your internal groups, even if the site is outsourced. Somebody has to be in charge right? This number reflects your group's total investment in Net-related activities.

Figuring out the return is a bit harder but still within reach. For example, the Net can be used to provide customer service. Generally speaking, as part of your customer service you have a phone center that takes calls, answers questions, and provides support. A cost is associated with each call that comes in, say, $10 per call. A certain number of calls come in each month, say 100, and if we multiply the number of calls by the cost of each call, we have the variable cost of the phone center. The total for this example, 100 calls times $10 per call, is $1,000. Assume that by placing customer service information on your Web site you measure 5 hits per month and 5 E-mails. Further, you record a drop of 5 calls in your phone center. You can make the correlation that the Web is now accounting for a 5 percent reduction in calls, resulting in direct savings of $50 per month or $600 per year (assuming no increase in either calls or hits to the Web site). The costs that you figured in developing that portion of the site totaled $300. Doing the ROI math, we put in an investment of $300 that saves us $600 for a return of 200 percent. Not a bad investment.

Adding to Return on Sales

In the return on investment example, we looked at how expenses can be justified through cost savings and that is good. Many of you are looking at the Web and thinking, "Great, I can save money but what I really want to do is to rationalize the Web as a way to make money"; then read on. Return on Sales is traditionally used to measure the performance of a given marketing campaign. Measuring your real results via the Web is surprisingly easy. You have dependable tools to measure the effectiveness of your campaign, as we have mentioned and will continue to mention throughout this book. If someone places an order through an order form on the Web, you can pretty much assume that they were lured into buying through your site. To measure the return, look at how much of your site is devoted to selling, for example, 50 percent, and take the total cost of the site and divide it by that portion; if the total cost is $1,000, then $500 is devoted to marketing and sales. Now we know how many orders we receive from the Web (that is the easy part); for this example, we will use 10 orders generating $5,500 in sales. Further assume that each of these widgets that we sell costs $100 to make. First, we need to determine total costs of what was sold. Add the cost of the portion of the site devoted to sales ($500) to the total cost of the items sold (10 items multiplied by the $100 cost per item for a total unit cost of $1,000) for a grand total cost of $1,500. To determine profit, we subtract our total cost from total sales, or $5,500 minus $1,500, for a profit of $4,000. Measuring the return is as simple as dividing our profits from our sales, or $4,000 divided by $5,500 for a whopping Return on Sales of 72.7 percent!

Making Customers and Suppliers Part of the Process

Why limit your engineering efforts to your company? A quick read in any management book of the 1990s says to incorporate your customers into the value chain. Customers are often disparate groups, ranging from the end user or consumer to your suppliers, service venders, and so on. The re-engineering question becomes, "How do I streamline these people into my organization without adding overhead to my flow of information?" or "How do I provide the U/A connection?" Consider, for example, the complexity associated with a large-scale retail operation. Selling goods to the public is the easy part for a company like Wal-Mart. The hard part lies in the logistics of keeping thousands of suppliers shipping, invoicing, paying in a way that is as efficient as possible.

Wal-Mart's Web site offers the first door for entry to many potential suppliers through its Retail Link. This service connects members of the Venders-Partners program through an extensive EDI (electronic data interchange) network, allowing Wal-Mart to place orders and venders to receive payments electronically. This network was an early source of strategic success for Wal-Mart, keeping costs down and ensuring a low carrying inventory for Wal-Mart's operations. The good folks at Harvard have written at least one case study of this aspect alone.

But what the Web site does is inform and set the tone for what is to come for a potential supplier. A folksy, down-home message from David Glass, Wal-Mart's president greets potential suppliers as they are led down the primrose path to supplier heaven. In their own way and using copy that is straightforward and simple, Wal-Mart sets a series of expectations that each supplier must meet. This no-nonsense approach is designed to weed out the two guys in a garage who think that if they can just sell to Wal-Mart their problems will be solved and provide Wal-Mart with partners who understand what will be expected before they start taking up Wal-Mart's time in the vendor process.

The Netscape Home Page

The growth of Netscape as both a company and as an Internet site has been nothing short of phenomenal (an understatement we're sure). The interesting thing about the people directing this rapid growth has been the continual returning to their Internet roots, meaning that they have not yet completely forgotten what has made them so great. The best way to get a glimpse into their thinking is to watch the evolution of their Web site, which is now on its umpteenth iteration (Figure 3–2).

But what caught our eye was a recent announcement asking for their customers to give them their impressions of the Web site and suggest ways to improve the layout and architecture. Imagine the effects internally if only 10 percent

Figure 3–2. One of the Netscape Home Page Incarnations.

of their 1 million plus hits a day sent in a suggestion or two. Surely this would have overwhelmed any support staff assigned the task of sorting through the plethora of incoming E-mail. But the upside in doing this is tremendous. Netscape opened its world and said, "Participate. Tell us what you want from us. Let us make our information, your information." That, in a nutshell, is what the Web is all about. Provide a way for the world to participate in a meaningful way if you are at all thinking about using the Web (internal or external) in a re-engineering effort. Sometimes the advice will be bad, no doubt about that, but one nugget gleaned from the manure pile and you might just have provided a better way of doing business.

METRICS AND MEASUREMENT, MEASUREMENTS AND METRICS

In a re-engineering effort, it sometimes becomes helpful to measure your accomplishments. This is what gives those involved in such efforts a chance at receiving a raise—which is the main goal of re-engineering after all. But the Web is new, the Net is elusive, and measuring success can be a painfully difficult experience. There is just nowhere to turn to tell if you are doing things right. At least, there was nowhere to turn until you started this chapter.

Measuring Net Throughput

Throughput, as we define it, is the average time needed to complete an on-line transaction. Throughput is not an absolute and single measurement, but rather a dynamic, changing variable measured on a variety of different levels. In the same vein, on-line transactions also vary greatly. In one case, it may be sending E-mail, another using file-transfer protocol (FTP) to send data, or accessing a Web page.

Because of the changing nature of the beast, what becomes important is the trend of throughput. We could, at this point, propose some advanced statistical measurement techniques a là Deming, but because this is not a stats book, we will stick to the concept of average throughput. So get out those stopwatches and time a few transactions. Let's define a few basic measurements:

E-Mail Throughput

- Send E-mail to someone in your department.
- Send E-mail to someone outside your department but within your company.
- Send E-mail to someone outside your company but in your country.
- Send E-mail to someone outside your company and in another country.

FTP Throughput

- Send a 100k file to someone in your department.
- Send a 100k file to someone outside your department but within your company.
- Send a 100k file to someone outside your company but in your country.
- Send a 100k file to someone outside your company and in another country.

Web Throughput

- Get a page from a server in your department.
- Get a page from a server outside your department but within your company.

- Get a page from a server outside your company but in your country.
- Get a page from a server outside your company and in another country.

Perform these tests at three or four different times on the same day and over a series of days. If you really have nothing better to do, try doing them every day for a month. Use the same servers or people to send files and E-mails to, so that some sense of a constant is maintained. The important thing is to remember that the times will vary greatly. This is not some empirical scientific test, be happy that you are conducting it, and have some fun with it.

Using the Results

Take the results and calculate the average for each category. For example, on any one day you add the four resulting times from sending E-mail and divide that total by four. The averages can then be plugged into a formula:

(Average E-Mail Throughput) + (Average FTP Throughput) + (Average Web Throughput)/3 + Average Net Throughput

Now take that result and plot a trend over the course of a week, a month, a quarter and see what happens to the graph. If your throughput increases, great! If your throughput decreases, then not so great and it becomes time to look for bottlenecks.

Bottlenecks

If it is the connection that counts—and as we mentioned, we believe it is—then it is the speed of the connection that defines the speed of the organization. It follows that anything that serves to slow down that connection effectively slows down the entire organization.

Some problems we have no control over, such as the resistance in a wire that slows down the speed of the electricity that carries our packets of information across the network. But in many cases, we can change or improve the processes and functions that cause not only our packets to slow down but also our personal interface with the network to slow down. In the following sections, we look at some of these common bottlenecks and describe ways to deal with them.

Bandwidth

The single biggest determinant for throughput and the cause of the greatest bottleneck is bandwidth. Think of a bandwidth as a pipe that carries water. The bigger the pipe, the more water can flow. The plumbing for a city block usually consists of a main pipe that feeds smaller pipes, and these in turn feed individual houses. It also follows that the more pipes that lead into and from the main pipe, the greater the demand on that main pipe. And this affects the pressure of the water in that pipe. More connections, less water pressure. It works the same way for computer networks. The main pipe is the line that connects to the Internet. The smaller lines are the connections of departments, business units, and other local or wide area networks into that main pipe. So the pressure in the pipe gets lessened as the number of connections increases. But as we mentioned earlier, it is *making the connection* that is important. When plumbers need more pressure, they increase the size of the main pipe. If your company needs more connections, you may need to increase the size of your main pipe. Another way plumbers increase pressure is by adding a second or a third large pipe to off-load capacity on the main pipe. If you think like a plumber, visualizing your network architecture becomes easier.

Proxies and Firewalls

Creating a secure internal network is of the highest priority to any company doing business on the Net. The idea is to keep people outside your organization off your internal network, but allow access to the Internet to the people within your company. The problem is that the inflow and the outflow occur over the same wire. To keep your network secure, some engineer in IS has more than likely placed a proxy server or a firewall on your network. By way of definition, a proxy server is a physical machine that acts as a single point of entry and egress to and from your internal network. All your E-mails get routed out through this machine, all your Web requests get routed out through this machine, all your FTP files get routed through this machine. This, needless to say, is one busy machine. The point in having one is to limit the number of connections to the external Internet. If your entire company has only one or two physical machine connections to the Internet, the number of potential break-in points is reduced.

A firewall usually works in conjunction with a proxy server. Firewalls are software solutions that monitor access through your proxy server and block out requests that are not approved. Combined, the proxy/firewall solution acts like a policeman patrolling a single point of entry to your network.

The other thing that this combination of security does is create a bottleneck for network traffic. Remember, every packet of information that you and your cohorts request has to travel first through a proxy server before hitting the Net. Not so bad if you have the network all to yourself, but if you are part of a large organization, your packets may be competing with the packets from every other person in the company. There is a physical limitation to the size of the pipe (bandwidth) so lots of those packets are either going to wait in line to get out to the Net, or are going to be rejected. Even those that are lucky enough to get through are still impacted. These survivors will be slowed down, since the pipe will be running at or near capacity and friction, that nasty old physics law, will occur.

Alleviating the strain of a proxy or firewall is more vexing than the bandwidth problem. With bandwidth, you can always get a bigger pipe. Firewalls and proxies are security measures; not implementing them leaves your internal network open to outside attacks that will surely come. The solution comes in increasing the number of connections, a theme in this chapter. Your IS people will complain loudly, but by increasing the number of connections to the outside world, you will reduce your Net throughput.

Design

You may not ever realize it, but the way you structure the design of your pages has a large impact on throughput. The reason is simple; the fewer clicks that a person needs to make to find information, the less time spent. For example, if your network serves a page in 10 seconds (a slow number but a nice round one for this example) and it takes a user six clicks to reach the page showing the information that is needed, then that user has spent 60 seconds navigating your structure to find the data. If anywhere along that line a click can be eliminated, you achieve a 10-second improvement in throughput. Consider Yahoo! as a site designed for speed in navigation.The value of Yahoo! lies in pointing people to other things, so technically the less time you spend finding what you want, the more time you will spend on Yahoo!, since information is so easy to find. A visit to the site reveals a very thin hierarchy, even though Yahoo! catalogs millions of sites. From their top-level listings down to the actual page pointers, we found an average of 5 clicks and, more often than not, you could bypass some of these 5 and get to information in 2 to 3 clicks. What's more, on the very first page and at the top of every lower level there appears a search box allowing the user to bypass the hierarchy completely. Each link within the site leads to something useful, either another set of links or pointers to sites. No baggage in the form of flowery text and graphics, just pure navigation.

In applying the lesson of Yahoo! to our own sites, three rules spring to mind:

1. Flatten the hierarchy.

2. Avoid extraneous graphics.

3. Make the links work.

The Net Itself

The great equalizer in the Net throughput story is the Internet itself. The Net is a cobbler's child and, like all cobbler's children, must eventually grow old enough to make its own shoes. Because the overall throughput of the Net is defined by its slowest link, somewhere between you and your Net destination there inevitably lies a 56k line. It will not matter how fast your new T3 is, if it encounters a 56k line the effect is like driving a Ferrari into city traffic: Your friends might be envious, but their cars go just as fast.

Up until a couple of years ago, the backbone of the Internet was under the control of the National Science Foundation (NSF)—you may have run across the term *NSF Backbone* somewhere in your readings. A backbone is a big pipe that runs down the city block (to go back to a plumbing metaphor) connecting all the buildings to each other. Commercial long distance carriers and ISPs like

MCI, AT&T, and UUNet have their own backbones that they are upgrading or putting into place to specifically handle Internet traffic. Again, because the Internet is a public network, the data carried across it are mingled with the telephone calls and faxes that make up the phone network. If you use a dial-up account, you may not have a separate phone line to handle the Internet. Yet. All of this acts to decrease the available bandwidth of the entire network. So the big players are trying to decouple the packet-based Internet information from the switched telephone network. The result should be a faster Net and one less bottleneck.

BENCHMARKING AND THE WEB

In the day-to-day grind of getting your work done, it is easy to become myopic. We know that this is particularly true of fast-growing industries where it seems that the world is waiting on your every move. It becomes all too easy to believe your own press. This represents a very dangerous state for corporations, since it is usually from left field that our competitors come.

A relatively easy way to test our own mettle is to benchmark our practices against our competition, or our processes against recognized industry leaders. For example, if you run a shipping department, you will want to benchmark against Federal Express and UPS since these companies set the standard for shipping packages. So how do you do it? Easy. Using the shipping example, both FedEx and UPS allow users to track packages, arrange for pickup, set up a new account, in short, a range of possibilities to make the user's life easier. How does your shipping department measure up? Can your people track their packages on the intranet? Can they arrange to have something picked up? The point is not to deride your shipping people, but to provide a useful benchmark for services.

In a broader sense, you can benchmark your entire Internet presence against your competition. Doing this is easy. Pick your top competitors and create a scorecard:

- They do/do not have a Web presence.
- They do/do not include detailed product information.
- There is/is not a way to make a purchase on-line.

The actual criteria will be up to you to determine, since you know what is important to measure in your industry. The purpose of the exercise is to determine where your competitors' competencies lie. Are they a generation ahead of you with on-line transaction processing, a secure, externally accessible extranet linking their suppliers, and real-time videoconferencing? Or are they stuck in the three-page, static HTML world that lists their press releases and a photo of the CEO?

Do not forget to run the scorecard against your own site. Be as critical, or more critical on your site as you are with the competition. When you are finished, combine the results and perform a side-by-side comparison. From this, you should be able to pick up the relative strengths and weaknesses in each site and see areas for improvement in your own.

The next step is to benchmark with sites outside your industry that are recognized as useful or interesting. From a few different lists, such as the Tenegra Awards for Internet Marketing, you can find sites that go beyond simply providing cool graphics to those that truly provide business value. For example, the AT&T 800-number service page gives you a way to search the 800-number directory for a telephone number. Good news for those who constantly misplace Post-it Notes! So in addition to providing ways to find out about AT&T and their goods and services, they publish useful information that anyone can use to find a telephone number. AT&T makes money when you call their 800-number information service, but they choose to provide value to their Internet offering with this service. We do not have the hard facts, but we also imagine that this saves AT&T some money by allowing those operators additional time to do other things or possibly by reducing the number of operators needed. How does that benchmark against your information? Let us say that you operate a call center that handles customer inquiries that can usually be answered quickly and without much technical detail. Is that information out on your Web site?

TEAM BUILDING

A considerable amount of time is spent in any re-engineering effort on making everyone feel comfortable with the process and in achieving project "buy-in." The process of building this consensus is known, for want of a better term, as team building. In the old days, team building was simple: The boss told you what to do. Now that we are all "empowered knowledge workers" seeking "self-realization and actualization through meaningful work-related activities," we can no longer be told what to do. We have to "want it." And as we look around for examples of who builds the best teams, we are ever increasingly drawn to the military.

A military battle team has to operate as a single, cohesive unit, capable of acting in complete unison to accomplish a single objective. The conditions of their employment, namely warfare, are harsh to put it mildly. Constant stress, lack of sleep, physical, mental, and emotional fatigue, as well as a host of other discomforts make team building difficult and necessary. Not that any branch of any military is necessarily more of a team than other branches, but the U.S. Marines may have the most unique approach to getting their collective act together. And how, you may ask, does this twenty-first century fighting force approach the daunting task of shaping young men and women into a rough and ready fighting force? Why they play Doom, of course . . .

Indeed, a visit to the Marines' Doom Home Page reveals a great deal about the way in which people are trained to act as a team. For the uninitiated, Doom is a multiplayer, "shoot 'em up" game designed to provide teenagers with an outlet for adolescent angst. The Marines, while certainly possessing their own brand of angst, found a much more useful purpose for the game than simply racking up deaths and endless mayhem. The Marines use Doom as a way to build teams. Programmers within the corps have modified the weapons in the game to reflect those that Marines normally carry (not that a few plasma deathray guns would be bad) and they created a realistic three-dimensional world in which to interact. This gives recruits a chance to fight it out together, at much less expense to both the government (in dollars) and the Marines (in lives) than live training exercises. Players can be killed or wounded so the Marines benefit by experiencing the real stress (these games definitely leave the palms sweaty) and consequences of acting as a team.

What the Marines learned is that with the help of a $30 game, they could simulate an environment that forced people to learn to work together. They also, judging by the attention to detail in the various levels and weapons, had a good deal of fun creating this world and participating in the action. Good cheap fun that helps people work together has to be one of the critical factors for any project's success.

ALIGNING BUSINESS PROCESSES

The Web can be the universal amalgamator, conglomerator, and universalizor. So much for the hype. In this section we move to align your Nets, both internal and external, into regular business processes in ways that will make them more efficient. We present a few different scenarios and look at the way to use the Net in each.

Field Sales and Demand Creation

Making a sale requires information. That blanket statement becomes self-evident when you consider the amount of time spent before making a major purchase. In fact, the time spent researching brands and products may be greater than the time actually spent in the final negotiation of a sale.

If information is vital to making a sale, then your salespeople should understand demand creation. In this sense, demand creation is the act of providing the information necessary to lead a customer to a buying decision. "But isn't that what advertising does?" you ask. In some cases, yes. But what you really want to do is not push someone into a sale (the advertising paradigm), but pull them into a sale by providing enough information for the customer to make an intelligent and informed decision.

The example for this is the Amazon.com bookstore. The reason behind Amazon's success is not necessarily their large inventory (which they have) or their fast service (1-week delivery) but rather the amount of information that Amazon provides. It is this information that creates demand. Amazon, if you have never visited the site, gives editorials, reviews, and even allows customers to post their own reviews and opinions on any book in print. The average Joe looking for a book has all the information needed to complete a purchase even if he does not know the book that he wants. Say he is vaguely interested in business. He can perform a search by topic and come up with books that range from accounting primers to advanced strategy management. He chooses a topic, say "Web strategy for competitive advantage," and refines his search to that subject. At this level, he is presented not only with a listing of available books, but also the opportunity to read what the authors might have to say, check out any relevant reviews, and even delve into what other people might want to share on the topic. In short, this Web site provides all the information needed to complete an informed transaction and, more importantly, the *demand* that was created because of that information.

Time to Market

One of the great things to come out of any re-engineering project is reduced time to market for new goods or updates to existing products. The reason for this is simply a reduced number of information points, which means that the information needed to make decisions flows faster to the people who need it, thereby reducing the total time needed to make and to implement a decision.

Hewlett Packard is one of the world's great manufacturing companies. If you are an engineer, chances are your first real calculator was an HP. They are also an extremely innovative company, producing among other things first-rate printers and computer peripherals. Since this is, after all, the information age, each printer must be supplied with software known as print drivers. These drivers allow a software program, such as a spreadsheet or CAD program, to talk to the printer so that the printer can be used to its full potential. Many of the printers that HP sells can do more than just print text; they can do colorful graphics, 3-dimensional art, and so on. Since software that is not made by HP needs to interact with an HP printer, and since that software is constantly being upgraded and changed, HP must, in turn, constantly upgrade and change its printer drivers. And since HP produces more than one model of printer, each model has its own unique set of drivers.

Lewis Platt, HP's CEO, was quoted in *BusinessWeek* as saying:

> Customers call us for new printer drivers two million times per month, costing us $14 a call. So we put a printer driver library at our Web site. It drew 1.2 million downloads in the first eight weeks and that number is rising 20 percent a month. We are saving $8 million a month.

In addition, these customers are getting the latest drivers without having to wait for HP to design packaging, document, duplicate floppy disks, package, and ship diskettes out to the waiting customers. HP has turned a process that took weeks into a process that takes days (see Figure 3-3).

The Free Market for Ideas

To paraphrase Adam Smith, the ideas that have the most merit will win out in a free, open, and competitive environment. Or at least that is the theory. Putting a free market for ideas into place can become a great competitive advantage for a company that can truly implement it. Some major economists (which we are not), argue that

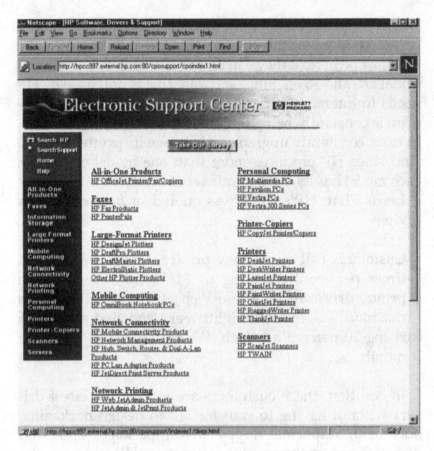

Figure 3–3. Electronic Support—HP Style.

what matters most in the economic growth cycle is original ideas. Yet corporate culture at most companies seems expressly designed to stifle the very things that could save them. A testament to the fact exists in the popularity of the comic strip *Dilbert*.

The trick then, in terms of re-engineering and moving toward a constantly improving organization, is to capture ideas and distribute them across the organization. The neat thing about ideas is that they often fall into generally definable categories. So ideas like, "Let's call our customers when their invoice is more than 90 days overdue" and "Let's not pay any of our invoices until they are at least 90 days overdue" fit into categories like "Finance" or "Accounting." And if we express our ideas in a way that lets other people know about them, we start to see our ideas catch on in other departments or divisions. "Hey, I hear that Division A doesn't pay their invoices before they are 90 days overdue. Let's try that."

Capturing and distributing information is the role of the re-wired company. Perhaps one of the most fascinating aspects of the Net occurs in the newsgroups. These threaded discussion groups allow users from around the world to chime in on topics that span the breadth of human experience, from animal husbandry to current events. Anyone with access to a news server and a news reader can take part in this great discussion. So how then do we incorporate that resource into the scheme of a re-wired organization? Simply start an internal news server and give people access.

One of the first things that went on-line in the General Electric intranet was a bulletin board to exchange the best practices in each of the various GE businesses. Topics were arranged by area of expertise. So manufacturing, procurement, and similar topics became the first areas for people throughout GE to contribute their ideas to. One concern was that the newsgroups would generate more questions than answers. To ensure that this would not happen, each topic had an area expert assigned to it. These experts were responsible for tracking down answers to questions, maintaining the quality of the discussion, and generally making sure the system was used. The outcome of setting this system up quickly exceeded expectations. Questions poured in to the system but more importantly, answers came in as well. And not all the answers came from the business process experts. The rank-and-file users with practical experience shared their views, solutions, and notes. What started as an exercise in collaboration soon became a viable and necessary way to spread information and allow shared access to knowledge.

Accounting

Of the two things in life that cannot be avoided, at least death doesn't involve accountants (at least for the departed). Now that we have made our obligatory accounting joke, we can get on with the story. Accountants and chief financial officers are finding that the demands of their jobs are significantly increasing in the on-line world, especially in small to medium-size businesses. Where a company could once get by selling into the domestic market, the Web has broken the boundaries and opened up the world to the around-the-clock global marketplace. But we already knew that. What we did not know was the complexity of keeping track of the tax implications of doing business in that global marketplace. Imagine you are the largest German exporter of marshmallows, looking at those kooky, sun-drenched Californians and saying, "What they really need is a' good marshmallow." In California, a marshmallow is entertainment, or at least taxed at the same rate as entertainment. So where's an exporter supposed to go for that kind of crazy information? Ernie, of course. Or rather, the Ernst and Young (E&Y) Web site called Ernie shown in Figure 3–4.

Ernie, in not so many words, is a bandwidth filler. E&Y gets paid for thinking and any minute that one of their consultants spends on the job not thinking is not just a minute lost, it means dollars lost as well. E&Y is not unique in this scenario; many consulting firms are up against the same dilemma. In fact, the major bottleneck in any thinking organization is a stoppage in the thought process. But the problem becomes, What if you have some free time and

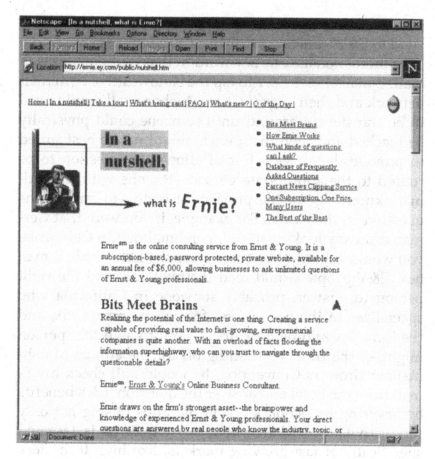

Figure 3–4. Ernie—Ernst & Young's Bandwidth Filler.

your clients' problems are, for the moment, solved? How can you constructively fill that free time without wasting time physically moving to another customer location where you could reasonably begin the billing process? And (there is always an and), keep the costs of providing the service low? The answer that E&Y came up with was an on-line answer center that tapped into a market that usually would not have the money to ask business-related questions of the people whose time E&Y wanted to fill. Ernie is a subscription-based service for customers who

pay a retainer that entitles them to E-mail questions related to E&Y's fields of expertise and receive answers from
their professionals. The key word in that last sentence is E-
mail. E-mail allows E&Y to tap the bandwidth of both their
network and their organization. A phone call would need
to be transferred around until someone could physically
be reached and then that person might not be best suited
to provide the answer. E-mail allows the question to be
routed to the appropriate person—the one with the intimate knowledge of the problem at hand—to answer the
question on their time. For example, if you were that German company looking to sell marshmallows in California,
you would submit your question to the local Ernie E-mail
box. E&Y people would read the mail and find the right
person to answer, probably someone in California who
specializes in the use of food for entertainment tax, and
the E-mail would be forwarded. Even though that person
might, at the time, be heading off for the annual Marshmallow Growers Convention, he would still check his E-
mail from the hotel and answer the question. E&Y benefits
by lowering its people's downtime, thus raising not only
productivity but profitability as well. Ernie is targeted
specifically at fast-growing markets, like high tech. E&Y
figures that these people are the ones most likely to become future E&Y customers after that IPO and are the
ones most in need of the kind of disparate help that Ernie
can provide. Not to mention that these folks already have
the basic network infrastructure that allows them to ask
and receive answers via E-mail.

Information Management

Perhaps no other area of operations is better suited to taking advantage of the Net than IS. Finding resources in this

area is particularly easy and any search engine will return lengthy lists of sites to visit. All the IS print magazines are represented; *Infoworld, PC Magazine, PC Week,* and so on. But as in other areas, the Net offers some unique solutions to help in everyday IS situations. We could fill a book on all the sites that offer some kind of software to download, patches for bugs and updates as well as competitive pricing information and sales brochures. But this book is about strategy, after all, so while those resources are out there for the taking, this section is limited to the solution of a common problem.

All too often, what IS managers face is a breaking crisis. Computers locking up, servers going down, applications crashing. What these managers need is answers and answers quickly. The first instinct has been to pick up the phone and dial a tech support number, placing yourself at the mercy of the great telephone lottery. Whoever calls first gets answered first, regardless of importance, and too often regardless of the provisions in your support contract.

Dave Winer, columnist for *Wired,* author of *DaveNet* and one of the last Macintosh software developers wrote:

Last week, in a postscript to "What's Going On?" I asked who did the song. Was it Smokey Robinson? No, of course not. It was Marvin Gaye. I'm lucky. In a moment of forgetfulness, when I want to know something, I can ask 40,000 people the same question. It was an easy one. Lots of DaveNetters are Motown fans. I got a lot of E-mail.[2]

That simple statement sums up the power of the Net. Answers in spades, 40,000 of them, to a single and simple question. That kind of power can easily be brought to bear using Usenet newsgroups and asking specific questions. The comp. Area of UseNet boasts over 700 different newsgroups ranging from comp.ai (artificial intelligence) to comp.windows, all with active and real people waiting to ask and answer your questions. You may not receive 40,000 replies, but you should get a few. In general, response time on the Net is one day turnaround; that is, your question will more than likely be answered the next day. Although this may not put out a fire, it may be more beneficial if the answer includes things like universal resource locaters (URLs) to sites where software fixes can be downloaded, contact information of people who have similar problems, or sample code embedded in the message. The first step is placing your problem on-line; chances are others have the same problem or have figured out a fix.

4

The Network Inside

In Chapter 2, we outlined the business models that work on the Net and the Web at large. We described attractors, push/pull models, and the like to better understand the wired world and our place within it. In this chapter, we hope to provide that same valuable insight in using Internet-based technologies, such as E-mail and the Web, inside your company on what has been dubbed an intranet.

Just what is this thing called *intranet*? Before we can answer that, we need to take a look at the flow of information within a typical company. Information is typically grouped into little islands or information archipelagos (collections of information islands). These islands have their own little cultures, people, and history. Sometimes the islands become big enough to become continents, especially if the information is vital to a company. What then, is lacking from this geographically diverse information system? To follow the metaphor a bit too far, what's missing is the globe. The Information Earth, in its entirety, should contain each piece and every bit of information lands within it.

That is what an intranet is—the globe. It is the connecting and underlying force that allows users to navigate from the Isle of Accounts Receivable to the Land of Benefits and Pension Plans. It enables communication, via E-mail or a Web browser, with the distant shores of the company's corporate world, and if well implemented, provides access to other worlds as well (via the spaceship Internet).

By now you may be thinking that we have spent too much time in California, but the theme is what is important: a connected and interconnected corporate world. So now that we know what it is and how to think about it, what can we do with it? We are going to look at intranets in two contexts: first the underlying things that make up an intranet, second an intranet as it applies to the good old value chain.

THIS THING CALLED INTRANET

We think that as important as the Internet is to doing business in the coming years, intranets may be more important still, with one very large differentiation. The Internet is where money will be made, the intranet is where money will be saved.

Again we ask, "What exactly defines an intranet?" Simple questions are never easy. Intranets are a collection of computer networks that interconnect within (and this word makes all the difference) an organization. Internet equals outside, intranet equals inside. At the infrastructure level there are many corollaries between the Internet and intranets. Intranets are made up of Web servers, news servers, E-mail servers, TCP/IP, routers, ATMs, and the like, just as the Internet is. If you are starting to feel confused; relax, we'll leave this to the techies and get back to business.

The way that we think about using an intranet becomes as important as, or more important than, what makes it up. We will stick to the three-perspective rule, since it lends a certain architecture to our thinking.

First, and perhaps most importantly, the intranet can be seen as a communications medium. People can interact, exchange messages, view information, and so forth with people throughout the entire organization. Along with this comes the idea of communicating to the outside world. You can open up your company to suppliers, customers, the public at large! The whole concept of intranet as a communications tool has great power. When we think about reducing cycle times, cutting red tape, or organizing group collaboration, nothing speeds these up like improved communications.

Second, it is an applications environment. The intranet is one big client/server computing paradigm, our Global IS out of Chapter 2, but limited to our organization. Clients exist in the form of Web browsers, E-mail programs, and news readers. Servers are just about anything anywhere, from legacy mainframes to data warehouses to spreadsheets on someone's desk. The underlying connection between all these things becomes the glue that binds the information together.

Third, it's an application unto itself. An applications environment allows other applications to be built on itself, but an application exhibits certain characteristics of its own. The intranet is more than just a platform and we should understand why.

INTRANETS ARE COMMUNICATION

Intranets are an explosive topic. Pick up any computing magazine and try to find an issue in the past year that doesn't discuss intranets ad nauseum. Magazines do this because hot topics sell more magazines; we discuss them because intranets are a fundamental tool in the new computing paradigm.

One of our clients told us that for any new document in their company, 4 percent of the requisite time is devoted to creating that document and 96 percent of the time is devoted to distributing it. If you, like us, doubt that this is true, stop for a second and review the process. It takes, let's say, about 20 minutes to type up a simple memo. Productive time, well spent. Now you have to pass along your wise thoughts to the world. So it's off to the copy machine. Five minutes. Across the building to the mailroom. Ten minutes. Put a copy of the memo in each of your team members' mailboxes. Ten minutes. Back to the office. Five minutes. Answer a call about the memo and give your reasons for not distributing a copy to the supervisor of Group B. Five minutes. Talk to a coworker who pops her head in to say that she lost her copy and can she have another. I think by now you get the point. The problem here is that distributing the document requires you to "push" it out to your coworkers. This push equates to real work and real time, something akin to 96 percent.

Wouldn't it be much better if you could type a memo and leave it somewhere for your coworkers to pick up? Or if you could send it out secure in the knowledge that they would receive it? What we are talking about is a "pull." Your coworkers pull your memo to their desktops, read it at their leisure, and act accordingly. No more "It's in a pile on my desk and I never saw it," because it's now on the network available 24 hours a day for each one of them.

The first concept to introduce is the idea of *enterprise computing*, and no, we don't mean the talking computer on the *Star Trek* spaceship. Enterprise computing means that all systems, applications, and, most importantly, information is available to everyone within an entire organization. Pretty powerful stuff and an elusive computing Holy Grail. Information available to anyone means that line workers can learn the cost structure of a product, my finance people can gain insight into the latest marketing campaign, and the CEO can pick up a clue as to what the organization is all about. This information has always existed but in little islands throughout the company (back to that bad analogy). Fiefdoms of knowledge, for the most part, since the person who owns the information controls the information. But if there is any lesson to learn about information control and the intranet, freedom of information wins. Give away more than you get and yea shall be rewarded!

So what makes up this wonderful creature, this thing called the intranet? In the beginning there is E-mail. Simple and elegant, coy and delightful. In touch with the organization and beyond to the very Net itself. But if you talk to IS managers the world over, they will tell you that corporate E-mail is too hard to set up, too hard to maintain, too everything that makes you not want to ever bring it up again. Hogwash. The problem is that most major corporations have bought into proprietary standards in their E-mail. Microsoft, Lotus, and others have not based their E-mail packages on basic, free Internet standards. They want your E-mail all to themselves and that is bad. As mentioned earlier, open standards are an essential requirement, and intranets depend on open standards. So have your IS manager download a free SMTP or POP3 server (better yet, get an IMAP4 server) and pick up a free copy of Eudora. You won't regret it. You do not even have to know what SMTP stands for—just make sure that your IS people are using open standards

E-mail based on SMTP, POP3, or IMAP4 (three common and popular Internet E-mail standards) to run their mail servers. Distribute some free copies of Eudora, the Qualcomm E-mail program that supports these open standards and start E-mailing. You won't regret it and your IS people will be bringing their skills up to date to compete in the new, open standards world.

E-mail opens up your world to simple point-to-point communications and more complex one-to-many relationships. Most of this book was written in E-mail messages, and lots of time across two continents. You can send a single copy of a message to a single person or send the same message to multiple people. This solves your push problem in distributing a memo. Because you can also receive E-mail from your coworkers, it solves their push problem as well.

The next big thing that you can do with E-mail is communicate to the outside world. As good as telephones are, they are dependent on an immutable law of physics: time. We know too many executives in this global marketplace who rise each day at 2 A.M. for a conference call to Japan, go back to bed only to get up again at 4:30 A.M. for a call to England. Given, there are times when it is absolutely essential to speak and hear a voice. For the routine things, E-mail is a godsend. I can type my thoughts, send them off overseas, and receive my replies on (and this is the important thing) *my time!* In countries that are developing a base of software engineers, development of applications can occur virtually around the clock and it is facilitated by E-mail. The scenario goes like this: My Wall Street firm is ready to shut down operations around 6 P.M. Eastern Standard Time. As a dedicated IS type person working for the firm, I check my reports, and notice that there is a small bug in one of the new trading programs. I type up an E-mail and dispatch it to the crew in India. Then I'm off to home and family. Over in India, the day is just beginning. The development team checks their E-mail, notices the bug report, Telnets into my computer in the United States—which because it is now off hours is sitting quietly—fixes the bug, and sends me E-mail to that effect. The next day I come in, check the E-mail, and feel good. The time saved and the productivity gained can add directly to my bottom line.

THE WEB AS AN APPLICATION

People who write about the Web think that by putting your Human Resource's material on-line you have developed a Web application. This section will not tell you how to put your HR material on-line. But what it will give you is a way to look at the Web from an application perspective, which we believe wins out long after your HR material has gone stale.

Once the world belonged to the mainframe, housed in glass and revered as a symbol of all that was good. Once the earth was flat, too. The idea that everything should be on a large computer, isolated in some central office, with limited access and limited applications has gone the way of the snail darter, protected but not in great abundance.

The Network Computer Device (NCD) is a scaled-down personal computer. The promise of this device is that it is capable of connecting to the Internet or intranet at a savings of thousands of dollars over buying a traditional personal computer. Proponents claim that the NCD will change forever the hegemony of the operating system, since an NCD derives its OS from the network itself, rather than from internal storage. Indeed, all the applications used would come from the network, along with any files that might be needed and any storage requirements. You may have all your wants and desires met with the network and an NCD. For some of you who have been around for longer than, say, 23 years on this planet, this may sound a good deal like mainframe computing. Indeed, many pundits have likened it to a mainframe and dumb terminals. What differentiates a NCD from a dumb terminal/mainframe setup is the network.

Client/Server computing is an 1980s' paradigm, usually involving some sort of large computer acting as a server and some PCs that connect to it acting as clients. People spend lots of time and money every year trying to make these behemoths provide simple solutions to everyday business problems. So why didn't client/server take the world by storm? The answer, in our view, is that underlying technology was too complex to provide the kind of quick and evolving solutions that a modern corporation needs. Too much time was spent developing overly sophisticated user interfaces when what the users wanted was fast access to practical information.

Lots of pundits claim that an intranet is simply an excuse by IS departments to hook a legacy mainframe up to a bunch of PCs. Just add water and Poof! Instant intranet. They couldn't be farther from the truth. Think back, if you're old enough, to the heyday of mainframe computing: glass-enclosed rooms populated by lots of very serious looking folks who often seemed to have difficulty getting the right data out of the computer for the right purpose. Sorry to those of you who used to live in those glass rooms. There are several problems with the mainframe or even the network centric approach that intranets specifically solve. The two most important ones are evolution and interconnection.

Evolution

If there is one resource that the modern corporation lacks it is time. The wiring of the world has drastically and dramatically shortened the life span of many a good piece of software. In the old days, you could think in terms of developing a piece of software in months or probably years. No one expected anything overnight and there are historic examples of application deadlines slipping by more than a single year.

No more. For an intranet, the order of the day is speed and it is unusual, rather than usual, for applications to take more than three months. On most of the intranets that we have looked at, new applications appear every day. Sure, they may lack professional polish and the latest in eye-catching graphics, but these applications fulfill the most basic rule of information, timeliness. We had a project manager at a major pharmaceutical company tell us that his sales force would never use a Web browser, that they needed lots of "eye candy" (fancy graphics) because they didn't have the attention span required to learn anything like the Internet. It's easy to guess what happened here. The salespeople, frustrated with IS, contracted with a local ISP to host their Web pages containing sales memos, pricing, and the like. Word quickly spread through the organization, and other departments soon had their very own Web servers located on myriad local ISPs. Corporate IS lost the information battle and complete control of the situation. Good for the sales people, not so good for the corporation. Why not good for the corporation as well? Sensitive material was now in the public domain available to the public at large, including the competition.

But the point of all this was speed. Applications change. Today's spreadsheet becomes tomorrow's accounting system module, which in turn becomes marketing information in a data warehouse. Enter the "thin client" (a.k.a. Web browser). This single program gives us the ability to develop many sophisticated applications without spending significant time writing code for the user interface. The browser is the interface! Using the thin client approach, it's

easy to change what the user sees and how the user inter-
acts with an application. The back end (the database, the
business rules, etc.) can also change without the end user
seeing what has happened or even being aware of the
change. Complete rewrites of large-scale applications can
be a thing of the past! Cast off your chains you legion of
corporate developers!

Interconnection

A mainframe is a mainframe. Simple idea and one that IBM
profited on for a very long time. But a mainframe was never
intended to talk to a PC. For those of you that remember (or,
God forbid still use) one of those green screened VT100
mainframe terminals you know that a VT100 terminal emulation program is hardly real interconnectedness. You cannot connect to anything other than the mainframe. But hook
a mainframe into an intranet, put a Web-based front end on
your application, and you have removed a huge stumbling
block (i.e., the VT100 terminal emulation program).

A big part of interconnection is letting the user have a
single face to information. We have all used different programs to accomplish different things. But computer applications should be like picking up a book; even though the
cover of a book changes, I can still read whatever is inside. A
Web browser goes a long way to accomplishing that single
look for a user. Sure the HR Web will be different from the
marketing Web, but all I need to access either is a simple
browser.

The other part of interconnection is connecting to the
outside world. An intranet does not stop at the firewall.
This simple statement seems to confuse many otherwise
intelligent and thoughtful people. One enormous benefit of
giving users access to an intranet is that they can connect

to the outside world. This may happen with E-mail or it may mean browsing some Web pages, but it has to happen. The reasons for doing so far outweigh the reasons against, but for the sake of argument let's look at why not to connect to the Internet.

Reason 1

People will spend all day surfing instead of working. Seems valid, doesn't it? If we let people loose on the Net, that's where they will stay. But wait, some of you are thinking, my people are already the understaffed, overworked, underpaid, over-stressed statistics that make up this nation's productivity gains. When the heck will they have time to surf for fun? If you weren't thinking that, you should be. In an era of downsizing, very few people in positions that require use of the Net will spend more than their first few hours randomly surfing for pleasure. There simply is not enough time in the day to spend visiting "Aunt May's Web of Discount Liquor."

Reason 2

It's too expensive to give everyone access. Perhaps you haven't read that this is the information age? Or that markets are becoming hypercompetitive? Imagine not making the investment. Your engineers are out of touch with some of the latest university findings. Your salespeople can't receive E-mail that could close the sale or further the relationship. Your support people do not have access to the latest technical specifications because that kind of information is reserved for the engineers and is not on the internal network. We could go on, but by now you get the point.

Reason 3

It is too hard to find things on the Net. Yes, it is. Maybe it is almost as hard as delving through reams of reports, stacks of industry rags, and mountains of files, but who said new information sources were supposed to be easy? Sure, it's a poor user interface and, yes, sometimes you do have to wade through lots of Yahoo! listings to find the URL with the SIC (Standard Industry Classification) code for corn farming but at least you can find it. In fact, once you have found it, it becomes a more powerful resource since you can save its location and refer back to the information whenever you desire, all without leaving your desk.

But connecting to the outside world is perhaps the biggest reason to start an intranet. Dave Winer, who writes for Wired and DaveNet (his own E-mail newsletter), speaks of the ultimate company being one that gives employees complete access to the Net and their own home page. No restrictions. As Dave would put it, *cool.* Think of the power in that simple idea. Think of the responsibility. The thought of workers, employees, managers, friends publishing whatever they want is scary to most managers. What would people do with it? Who knows. But maybe, just maybe, people would use it for the tool that it could be. They might put up information that really mattered, that would improve the bottom line. Or they might put up pictures of their children's first communion. The point is that they could. Peruse in a free moment the home pages of the various employees of Netscape and you will see what we mean. Their tastes range from homebrew (homemade beer) to technical descriptions of a T3 connection. What the employees get is an outlet. What Netscape gets is creativity.

The New Paradigm

So if client/server is old hat, and evolution and interconnectedness are what really count, what can we use as the buzzword for the new computing paradigm? We propose

Client/Net computing. Remember that client/server computing generally dealt with a single-purpose client connecting to a single-purpose server or in the best case, a single-purpose series of servers. Client/Net computing connects a multitude of clients to a multitude of servers, using either the Internet or an intranet. We already see this being implemented, with Web browsers in particular supporting a variety of open protocols. Using my copy of Netscape Navigator, I can view a Web page using HTTP, move files with FTP, connect to a news server, send E-mail. One browser and multiple servers enable me to complete multiple tasks. In the future, I will be able to connect my spreadsheet to the Net, my word processor, and my calendar—multiple clients talking to multiple servers to gather, maintain, and update my personal information.

One of the exciting developments in Internet technology is the advent of push distribution mechanisms. We have talked mostly about the Net as a pull: You pull down the information that you want. But companies like PointCast and Marimba are using the Net in the same way that television broadcasters use their networks, as a means to put information passively in front of users. Now before you start thinking that this is about entertainment and a TV on every desktop, think again.

PointCast is a screen saver and information tool, designed to take advantage of the time that you spend away from doing work on your computer. The software connects to a server on the Net that pushes information, like stock quotes and breaking headlines, to your desktop. Now instead of watching flying toasters, you can keep track of your portfolio. The advantage to the intranet is that this technology is customizable and you can connect it to an internal server, rather than the PointCast external server. So you could conceivably run training sessions, show current sales figures, or broadcast inspirational messages.

Marimba is specifically designed to help keep network software up to date. The coming of network applications that are based on something like Java or ActiveX technologies will no longer reside on your computer, they will reside on the network. The versions that you load in the morning may change during the day as bugs are fixed and new features added. So the Marimba software is designed to keep your desktop up to date with the network, seamlessly. No more diskettes to load, CDs to break, or some technician ripping apart your computer saying, "Now where is that hard drive again?"

For years now, users of good old E-mail have had wonder and robust push technology. Those of you without the means or desire to spend hard-earned dollars on the latest in push technologies, can pick up a copy of LISTSERV, which is a mailing list management system designed for E-mail. The LISTSERV software allows you to create and maintain a list of recipients, usually who are centered around a particular topic or are all part of the same team. LISTSERV allows members of the list to broadcast, or push, messages to the entire list even if they do not know all the E-mail addresses of the recipients. It acts as a forwarding device, putting your E-mail into their boxes and their messages into yours.

CERT, the Computer Emergency Response Team, maintains one of the most active and valuable mailing lists.

THE INTRANET AS AN APPLICATION PLATFORM

As we have described, the Web as an application, is an end unto itself. But an intranet can also be used like a computer operating system (Win95, MacOS, etc.) to build applications.

A Net opens up applications from limited geography. By that, we mean two things: First, the programmers no longer have to physically be together to develop an application, and second, the applications need no longer depend on the physical makeup of the Local Area Network (LAN) to reach a user.

We promised to stay away from overly technical discussions and stick instead to management and strategy. So if you started reading this thinking that you were about to get a discourse on the C programming language and how it interfaces with TCP/IP, you are luckily wrong.

Strategically, some of the best people in your organization may be physically unconnected, especially as more of you compete globally. Centers of knowledge may be springing up in all the corners of the organization. Collaboration is an important part of any application development process and using an intranet (and the Internet) may be one of the most effective ways to bring these knowledge bases together.

There are some great examples on the Net of *virtual* teams working together to solve large-scale problems. The development of Linux may be the best collaborative effort thus far. Perhaps the thorniest problem in computing is the development of an operating system. Things like Windows, MacOS, or UNIX are all very large-scale development efforts involving big teams and long periods of time. Most are born from commercial need and are heavily financed endeavors. Most, but not all. Linus Torvalds, working at the University of Helsinki in Finland, wanted a UNIX-like operating system that would work on PCs instead of expensive workstations. His project requirements were fairly simple, the operating system would be distributed free of charge and anyone could modify it. But if you did modify it, you were asked to post your modifications so that others could incorporate them into their work. Linux, the name of the new operating system, was posted on the Net for downloading, and Linus let people know about it in a newsgroup posting in October 1991.

The response was tremendous. Developers across the Net (i.e., around the world) joined in the fun, writing everything from device drivers to porting Linux to new platforms. The features that found their way into the OS rival anything on the market and are as up-to-date as any existing commercially available operating system. What Linus gained in developing the system on the Net was access to more and varied programmers than he ever could have found by himself.

This is not to say that you should open up your internal development projects to the Net in hopes of getting the same response, but you should seriously consider creating a geographically disparate development team. The tools are available; all that you need to do is implement the strategy. Open up development to anyone in the organization who is available to work on it. Be creative in your talent sourcing and realize that your intranet gives you access in ways that you previously did not have. Pick the best people rather than the closest people. What you may gain could be a more robust application and a better way of doing business.

THE INTRANET AS BEST SUPPORTING ACTOR

Now that we have covered some of the conceptual basics of what an intranet is, or might be, we should look closely at how an intranet works within an organizational structure. By applying our basic rules (Web as communications, applications, and application environment), we can look at the ways an intranet has an impact at every level of internal operation.

The Primary Activities of any business are defined in Porter's value chain as Inbound Logistics, Operations, Outbound Logistics, Marketing and Sales, and Service. Along with each of these primary activities comes a related Support Activity defined as Firm Infrastructure, Human Resource Management, Technology Development and Procurement. It is in these Support Activities when an intranet has the greatest value. Like the supporting actor in any good movie, the implementation of good support is the difference between success and failure. A good salesperson, who cannot get an order processed will soon stop getting new orders.

The bias in using a Net-based infrastructure to support primary activities is cultural. In numerous cases we've studied, companies have all the appropriate technology but the people are not using it. To be sure, E-mail gets sent around for birthday parties, wedding announcements, and so forth, but is E-mail linked to inventory so that a message gets sent to procurement when supplies run low?

Sales support is an easy and obvious example and one that affects most companies. Everybody has some sort of sales and sales support organization. Increasingly, salespeople work remotely, close to the customers. In some cases, these salespeople are physically colocated at the customer site. This makes for great customer support and should lead to increased sales and sales productivity. But it also makes for a sales communications nightmare.

Xerox invented many of the technologies that are in use on the Net today. Things like Ethernet, the graphical user interface, a mouse and a host of new applications that are coming out of Xerox's Palo Alto Research Center (PARC) show that Xerox understands the force of networking technologies. One of their recent advertisements implores users to "Network the Document." Paul Allaire, CEO of Xerox, has even declared that Xerox is to become the Web Document Company. Based on recent announcements and some interviews that we conducted, we think that Mr. Allaire's vision may actually become the Xerox reality.

But Xerox has some room to grow. For example, in doing research for this book and in providing some consulting for Xerox, we were never able to send E-mail to anyone inside the company. Every piece of mail bounced. The effect on our operation was minimal, since we had direct access to the people we needed via the phone or in person. But imagine an example where a valued customer needed to urgently contact a sales rep and had only an E-mail address. Apparently, we were not the only people who had this problem. At the source is the way in which Xerox routes its E-mail. It seems that nearly all Xerox's mail goes through a machine in Palo Alto, home of Xerox PARC. There are valid security reasons for doing this; a single machine routing mail does not give away the internal IP addresses of the machines from whence the E-mail came. But the business effect is disastrous. Xerox has a reputation built on providing the highest level of service in their industry. This confidence, which is so hard to gain, can be so quickly lost because of a bounced E-mail.

The moral is easy to understand, connected means both ways and it is up to your company to make sure that the connect is real.

The Web as Logistical Support

UPS and FedEx are engaged in a war of the Web: who has the more advanced site, who provides the most service, who gives you more information. While many industries have competitors outdoing themselves over their sites, few have done so in a more public eye and in a more studied way.

Logistics is one of the primary activities on the value chain, whether it involves moving goods within the company or moving goods outside the company. In the simplest sense, UPS and FedEx are logistics companies. Most corporations, large and small, do not have the extensive mail room facilities that once provided the future CEO with an entry level position. Shipping goods in a global business environment requires specialized and efficient logistical support.

So what value does the Web provide these two players and how does it relate to my intranet? A recent 30-second UPS television commercial that cost who knows how much money to produce and air, promoted the fact that UPS can track your package via the Internet. They will give you the information that you need, regardless of where in the world you are connected. Sure it provides a valuable support service; I can now find out the status of my delivery without the need of specialized UPS software or calling and waiting on hold, but that is not all.

What does the Net provide UPS or FedEx? They get to colocate their information onto your network. Think about the flow of data in a logistics system: You initiate the movement of goods, contact a carrier, transfer your information to their system, both of you track the package to its destination. The result? Duplication of information. Who owns the information on where the package moves? You both do. UPS effectively maintains your information on their network. By linking the two networks via the Internet and your intranet, you can access Internet information about your package on their network. The advantage to you is lower cost of information. The advantage to UPS and FedEx is improved service.

The Web as Service Support

The average sale of any good increasingly depends on information about the product itself, about the possible applications of the goods, and about pricing, inventory, and availability. Increased demands on information means that fewer people will have complete and intimate knowledge on all aspects of the goods that they sell. And as companies downsize, fewer people will be required to provide customers with more information. Phew!

We see an intranet as vital to providing information to the customer, both as an external marketing vehicle (in the case of an external Web presence) and as a way of supporting the sale or the goods once the sale is made. Often an external Web site is one of the most complete and thorough knowledge bases that a company possesses about its goods and services. Many times, we are finding that the Web site has better information than the call center or salespeople have access to from other methods.

Such was the case at Smith and Nephew Dyonics (now Smith and Nephew Endoscopy), a British maker of surgical equipment. Salespeople within the organization are highly distributed and in close contact with their customers (hospitals). Ordering of product is done either through the salespeople, in the case of new accounts, or via a telephone support staff for established customers. A typical sales call to the phone support would go something like this: Phone rings, person answers, and a question is asked. This usually results in the telesupport person putting the customer on hold, getting up from her chair, and leafing through a pile of material in one of Smith and Nephew's product shelves. Not very efficient. So Smith and Nephew used an internal Web server to serve up product information straight to the desktops of their teleservice people. This small effort has produced tremendous gains in time saved and better service to the customer.

Novell's business was built on its reseller channel. For the first time, networking PCs together in Local Area Networks became easy enough for the average or above average computer technician to architect. Novell broke out of the ranks of working through large-scale systems integrators to begin working with smaller, sometimes one-person operations. These resellers, in turn, focused on selling LANs to every facet of business, from small mom-and-pop operations through workgroups and departments in major corporations. Novell's end user who needed support was rarely the person sitting in front of a desktop PC in the marketing department of a Fortune 500; instead, it was the technician or help desk that supported that PC.

Providing technical support to this geographically and technically diverse group was one of Novell's biggest implementation hurdles. Providing phone-in support through an 800-number was provided but costly. Technicians were often called in when a network was no longer functioning at any hour of the day rather than the typical 9 to 5 hours of the phone bank technical support.

Novell, the company that popularized the Local Area Network (LAN), has long been an advocate of on-line technical support. Beginning on CompuServe, the NetWire Forum, sponsored by Novell on-line, became the preferred way of receiving technical support for many companies that needed advice in purchasing, installing, and supporting Novell networks.

In the early 1980s, Novell started a support forum on CompuServe called NetWire, a play on the name of their networking product, NetWare. NetWire promised that a team of technical support people would constantly monitor the forum and answer questions as the questions were posted. The overriding reason to provide the service was that the support team would no longer be limited and constrained by time. Phone support with an 800-number costs real dollars for every 6 seconds spent on the phone line. Answers to customers tended to be brief or unanswered with promises that the customer would be called back. As anyone who has spent any time on a customer support line knows, you rarely if ever get that call back. On-line support is another matter altogether. Answers can be as long and technically detailed as required. Supporting documents, such as technical specifications and white papers can be sent along with the reply. Typical questions can be gathered and answered in an FAQ (Frequently Asked Questions) document. And on good days, people who frequented the forum would answer each other's questions, post tips and tricks, software that they had developed to solve a particular problem, and so on. The end result was that the user received more support, from more places than would typically be available.

The forum was not open to the general CompuServe public since you had to register with Novell to join. This simple act meant that the average Joe would not clutter the forum with unnecessary and perhaps inappropriate questions. On any typical day, a user would find hundreds of posts with questions ranging from the simple to the overly complex. Each post had a response generally within 24 hours.

In 1994, with the explosion of the Internet, Novell moved the NetWire forum out of CompuServe and onto a Network News Server that Novell hosted. The effect was that now Novell could reach users that were not CompuServe subscribers. They could also still reach the CompuServe people, since these users had access to the Net thru CompuServe. Additionally, Novell saved the added costs of paying for their CompuServe forum. The move resulted in increased participation, better services, and reduced costs, not bad eh?

The Novell site (see Figure 4–1) offers another take on the colocation principle: moving the information from Novell's network onto the network of their customers, resellers, and integrators. By opening up their information to their public, Novell reduces support costs in other areas (like 800-number telephone service).

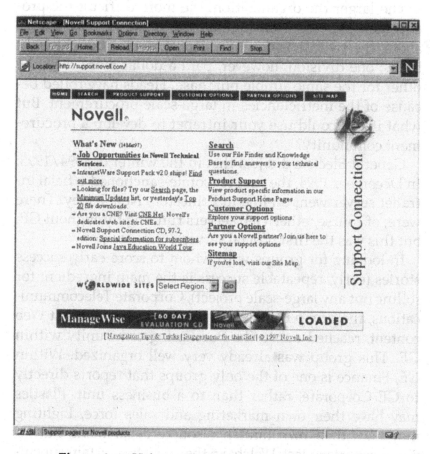

Figure 4–1. Helping Those Who Help Themselves— Novell Corporation.

The Web in Procurement Support

What is the biggest barrier in finding the lowest cost, highest quality supplier of a particular good or service? Information. The business environment today means that I can receive goods overnight from practically any spot on the globe. If I can find an efficient producer of golf balls in Madagascar then I should be buying my golf balls in Madagascar. But how can I find out if there are golf balls for sale in Madagascar?

The larger the organization, the more difficult the procurement process. Nothing new there. We all have stories about the two divisions of a company that bought light bulbs; one division, however, paid a dollar more than the other for the same simple purchase. Heads have rolled because of the inefficiencies in large-scale procurement. But what if you could use your intranet to develop a procurement community?

General Electric did just that in the winter of 1994/1995. In December 1994, the first corporate-sanctioned official intranet server went on-line in Princeton, New Jersey. There were, of course, other Web servers running throughout GE, but this was the first official server.

In looking for groups with whom to score early success stories (early, repeatable success is the main ingredient for rolling out any large-scale project), Corporate Telecommunications, the group responsible for developing the first Web content, reached out to the Purchasing community within GE. This group was already very well organized. Within GE, Finance is one of the only groups that reports directly to GE Corporate, rather than to a business unit. Plastics may have their own marketing and sales force, Lighting may have their own human resource department, but Finance reports to Jack Welch. So there exists a certain subculture within the finance community, a special camaraderie among the coworkers not unlike that found in any other group of coworkers. The difference in this community is that this team was geographically dispersed. People in Europe working with people in America. People in Asia negotiating deals that affected people in South America, and so on. Much of the decision making for routine deals was done locally. For example, a purchaser in GE Lighting might buy paper from a local vender for the local rate. The local vender, a small business operator, marks the paper up knowing it will take 45 days to receive payment. However, GE Corporate may buy paper from an internally approved vender with whom GE has negotiated a price lower than any particular business unit can negotiate on its own, due to the volume of paper involved. But GE Lighting may never find out about this agreement. This simple case, multiplied by the thousands of different products that GE buys from suppliers can make for huge amounts of overpaying based on lack of information.

So the very first thing that the finance community put online was all the corporate purchasing agreements. These documents were then indexed and plugged into a full-text search engine so that anyone could type in the product that they needed and come back with a signed agreement that gave GE the best possible price on any particular good. Other businesses soon realized the power that this afforded them and they too began to put their agreements on-line. Since GE often negotiates global agreements, now a GE purchaser anywhere in the world could find the best price from a GE-approved supplier anywhere in the world. Powerful stuff.

Most companies have spending limits imposed by their financial advisors on how much they can spend with a particular vender, either in a single purchase or over the course of the year. The idea is that, by limiting spending, you create a more diverse supplier base and less dependence on any one supplier of a good. The problem is that the supplier base can quickly become exhausted.

Again, because it is one of the world's largest companies, GE needs to expand its supplier base. Because they buy so much and so many products and services, many venders do not realize that GE may or may not buy their particular service. To expand their reach, GE rolled out the GE Trading Process Network. The GE TPN is a collection of tools aimed at the GE supplier community but also at the purchasing community within GE. The first part of the network that went on-line linked a database of SIC codes with a historical database of GE purchases. A supplier of sheet metal, for example, could enter in their SIC code and find out that GE buys $50 to $75 million worth of sheet metal across nine business units. It then gives ways to join the GE supplier lists and contact the various business units that buy those goods.

On the internal side, the GE purchasing community can see where their dollars go. They have a chance to determine the types and kinds of suppliers that are filling their need for materials. Again, they have access to a new pool of potential suppliers and can broaden their base of support. A side benefit is that GE business units can find other opportunities to sell their goods within GE, but according to Jack Welch only if they can offer the goods at a competitive price.

INSIDE OUT—LINKING YOUR INTRANET TO THE WORLD

An intranet exists inside your company, or at least that has been the focus of this chapter. In fact, keeping your intranet from the outside world is core to your business, since vital information resides there. No one would want a competitor to see their financial information, for example. But what may not be so obvious is the need for a link to the outside world.

The Commerce Community—Commerce.Net and Industry.Net

One of the first projects in large-scale electronic community building for the specific purpose of creating an information marketplace on the Web was Commerce.Net.

Begun in April 1994, Commerce.Net, led by early Internetworking companies such as Silicon Graphics, Sun Microsystem, and Oracle, is an attempt to link suppliers of computer hardware (e.g., chips, memory, circuit boards), with consumers of the same. Your normal supplier of logic chips is running low? No problem, hop onto Commerce.Net and see who has some for sale. The specific initial goal was to publish inventory information exclusively for companies in Silicon Valley, California. To ensure that companies that might normally be highly competitive would cooperate, a consortium was formed to run the marketspace and technology was provided by Enterprise Integration Technology, a small Internetworking systems integrator that was a leader in forward thinking Net-based commerce.

The initial pass of Commerce.Net exceeded expectations. Companies joined not just from Silicon Valley, but from around the United States and later, overseas. The initial focus of providing inventory information gave way to providing solutions in electronic commerce and Net-based transactions. The experiment continues, although EIT was recently purchased by Verifone (the credit card verification people) so work has slowed in providing commercial products. Commerce.Net works because it provides something that everyone in procurement needs: Information. Today, in addition to listing organizations and linking suppliers, Commerce.Net provides valuable information on Net usage and free software, and spearheads work in electronic commerce standards.

Industry.Net had a completely different industry focus but provided much of the same thinking that went into the original Commerce.Net. In 1994, many manufacturers in basic, noncomputer industries either did not think that the Net was for them or did not think about the Net at all. These manufacturers had other, more important things to think about than how to create a Web site. Industry.Net knew this and basically said, "Give us the information on the products that you want to sell and we'll take care of the Web site and the transactions, and it won't cost you an arm and a leg, and the risk is ours." This basic premise worked extremely well. At a time when most Web developers were concentrating on developing on-line shopping malls to compete with the Internet Shopping Network, Industry.Net took a business-to-business approach. Companies looking for goods to buy had a single source to turn to and a forum in which to conduct business. Manufacturers had a place to list their goods in a forum where buyers and sellers could meet. The big advantage in using this in procurement was the ease of use. Many pundits claim that while there is lots of good information on the Net, finding it is too hard and wastes too much time. Problem solved.

HUMAN RESOURCE MANAGEMENT

One of the daunting tasks in HR is human procurement, or finding the best people to fill a position (and if possible at the lowest price). That may not sound particularly pretty, but it should ring true. As in the procurement of goods, the procurement of people relies on information. Information about who is looking for work, what jobs are available for those people, their geographic location, how to contact those people, and how to start the hiring process. This is true whether the qualified person is within the company or outside of the company.

The typical MBA is hired by a company that visits the candidate's campus, conducts an interview, arranges the obligatory company visit, and so on. If you want a Harvard grad, you visit Harvard. But what about companies that (for whatever reason) want to extend their search beyond the capability of their recruitment teams? Certainly, there are qualified, bright, and energetic people who do not go to Harvard. The simple answer is to post jobs on your company's Web site. But what about the graduates who may never have heard that your company exists? Like selling a product, selling a job requires more than just posting some information. Literally thousands of resumes are put on-line by students each year and finding the jewel for your organization is a daunting task. The problem, again, is too much information and too little organization.

Prior to the Web, people either advertising a job or needing a position would post their request in one of a variety of newsgroups. Although many a good consulting gig has been gotten from the Net, to a corporate recruiter, this method of hit-or-miss hiring lacked a certain, shall we say, polish.

But then the Web came along and with it a start-up called CareerMosaic. CareerMosaic differentiates itself from the competition by focusing on the Human Resource department and providing a professional with the tools needed to get the best people available. A quick overview of their site shows areas for posting resumes, reviewing specific companies that use the CareerMosaic Listing Service, an industry-specific area for healthcare-related jobs and a resource center that provides information for people looking for work.

Delving into one of the company areas shows that CareerMosaic is smart enough to give employers their own space to create what they will. Employment opportunities and job listings are controlled by the companies themselves as well as the look and feel of the pages. This means that your company can use CareerMosaic to steer people to you, while controlling the critical aspects of presentation yourself, like the demand creation model mentioned in Chapter 3. No listing service like CareerMosaic is worth the time unless people on both ends of the transaction participate. You need to know that if you spend time there that you will be looking through a qualified applicant pool. Conversely, if you place your resume on-line or are otherwise looking for gainful employment, you want to make sure that the companies you want to work for are adequately represented. This is the advantage of the service over traditional print classifieds. Printed materials, such as newspapers, have a local to regional focus. It is up to the applicant to pull print information by subscribing to a newspaper from another area. While this is not a particularly hard thing to do, it takes some effort and some money. CareerMosaic, on the other hand, knows no geographic region nor is it a time-dependent medium like print. Want to post your resume today? Great. Want to advertise for that job now? Wonderful. Imagine the scenario when you can advertise, review resumes of applicants, and arrange for the first round of interviews all in the span of a single day.

Who's in Charge

In Chapter 8, we discuss the need to establish the Webmaster. The same holds true for the intranet, you will have to place someone in charge. In this case, however, you may already have the group you need.

The PC revolution was supposed to break the glass tower that corporate IS had become during the mainframe era. PCs on the desktop removed the central authority of IS and changed their mission considerably. The Net, even with its server centric ideals, does little to change this. Here's why; the network depends on connections (as pointed out in Chapter 3) not on a single connection (as in the mainframe world). Because of the need to connect, servers act as gateways and navigation points to information that ultimately still is controlled by the user.

One of the main new jobs of the modern IS department is to maintain the infrastructure. If we think about the types of things that the intranet will be used for—applications, application platforms, communications—then the job of keeping the connection alive is one of the most important in the company. Information Systems is already familiar with your infrastructure, trust them with the rest.

That said, make sure that everyone understands the role of the intranet and has the access that they need. Some groups definitely need to publish information and that information may need to change daily, maybe hourly. Sun Microsystems has developed some basic guidelines that they use to ensure that everyone has proper access and knows how to publish.[1] They have even opened up their Style Guide (Figure 4–2) to the public, to help others learn how to create Web pages. A quick look reveals subjects ranging from Audience, Security, and Page Length to Selling, Language, and Graphics. Each of the sections is laid out in an easy-to-read format, and since it is on the Web, anyone in the organization can access it at any time. The Security section is particularly interesting since it points out the cardinal rules of Web publishing:

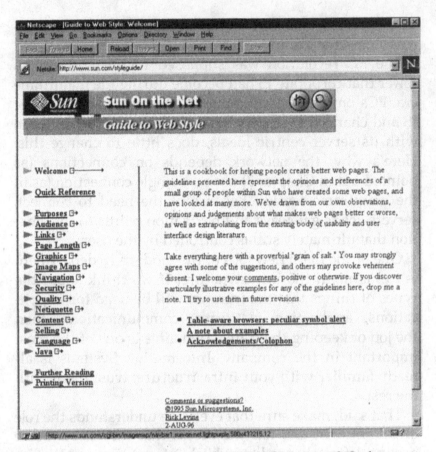

Figure 4–2. The Network Is the Computer, with Style.

Don't publish "registered" information.

and

Think twice about publishing "need to know" sorts of documents.

These two rules will go a long way toward keeping your information secure, especially if your intranet gets extended out into the Net itself.

Myopia

Regardless of your individual perspective on the Web, whether it be information systems, marketing, human resources, or whatever, one thing to avoid is thinking only in terms of your own needs. The Net becomes many things to many different people. It is all too easy to believe the press, the venders, or even the people in your own departments when they start talking about the be-all end-all that an intranet or the Internet promises. Beware the ides of March was the warning given Caesar, and look where ignoring that got him.

5

Marketing in the Marketspace

Since the explosion of the Web as a business medium, one of its primary uses has been for marketing. Corporate marketing departments see the Web as a deep advertising medium, where extensive information can be used to back up other advertising efforts.

In this chapter, we take a more reflective view of marketing. We consider where the Net and Web fit in by investigating how consumers respond to advertising and actually make purchasing decisions. We look at virtual advertising and the ways in which the Net and Web can be used to learn about markets, including the mine of data available in Usenet newsgroups.

CONSUMER BEHAVIOR AND THE NET

We will begin our discussion with a very simplified version of the common buyer behavior model (Figure 5–1).

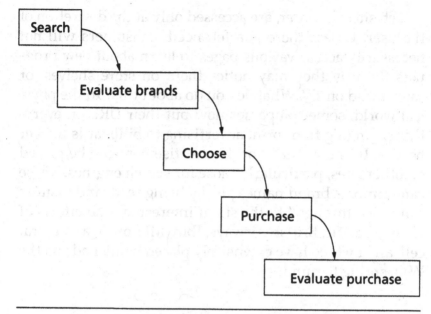

Figure 5–1. Very Simplified Consumer Behavior.

Consumer need can be stimulated in several ways. Consumers search out information on alternative products or services, evaluate competitive brands using some decision rule, choose one, make the purchase, and then evaluate it for future purchase decisions.

Need Recognition

Typical sources of stimulation include advertising, point of purchase displays, and promotions. Corporate control over these is considerable. Television commercials, for example, are often aired at the discretion of the marketer, not at the discretion of the viewer. Home shopping TV stations currently choose the order in which they present products for viewers to see. Catalogs present varieties of products on the same page to encourage crossover purchases.

Web sites, however, are accessed only at the discretion of the user. Unless there is a felt need, consumers will not necessarily access various pages to learn about new products the way they may notice them on store shelves, or catch an ad on TV. What do you do about this? In the physical world, some companies now put their URL on everything, ranging from print advertising to billboards to beer bottles. In the virtual world, advertisement can be placed on other sites, particularly those for search engines. These can reinforce brand names just by being read, and visitors can "click through" to the site if interested. The efforts of Cathay Pacific, British Airways, IBM, Microsoft, and Duracell, all of which have extensively placed banner ads on the Web, spring to mind.

Information Search

Consumers exert varying degrees of energy to seek out and process information as they learn about available products. For items requiring limited problem solving or search time, consumers use *proxies* of information (brand name, price, or past purchase history) as heuristics in their decision-making process.

Potentially, the Web makes the information search portion of the decision-making process much easier. In contrast to the time and effort required to go to many physically located stores or phone around, visiting multiple sites on the Web requires minimal effort (although at present, download times can make this frustrating). Directories and search engines allow users to browse through comprehensive lists of vendors arranged by product and service, or to search for a vendor by name or page content. This nonlinear search method provides unlimited freedom of choice and greater control for the consumer, contrasted with the restrictive navigation options available in traditional media, such as television or print. And all from the convenience of a home computer.

On the Web, comparative shopping on price and product attributes becomes even easier, as all alternatives are presented in proximity to one another. "Proximity" now means classification in a directory or the results from a search engine. Perhaps most importantly, electronic agents may even eliminate the need for searching. Consumers are not idle receptors; armed with agents and search engines (and who knows what in the future), they have the ability to search for products using specified criteria.

Evaluation

When dealing with information overload, a low risk product, or too many brands to evaluate, consumers normally use heuristics to inform choice. One argument suggests that the importance of brands may greatly diminish, especially in a world populated by agents. Decisions will be based on the actual delivered value. Loyalty to a brand may be established based on delivery of value, but will be subject to reevaluation as alternatives are introduced and are evaluated in a cost-free environment. An alternative argument, however, suggests that brands will become more important. Due to the huge increase in products and outlets afforded by the Net, confused consumers will retreat to known brands. These arguments are summarized in Table 5–1.

Trial is an important part of most evaluation processes and reduces the consumer's risk. While the success of catalog orders, home shopping, and direct-mail purchases

Table 5-1. Brands: More or less important?

Brands will become *more* important because . . .	Brands will become *less* important because . . .
Consumers will be confused by the increase in available products and outlets. They need heuristics to help them with information overload—and will use brands as a surrogate for quality.	Agents will search out products based on criteria. Consumers will be less influenced by image. Consumers will talk to many others electronically—the amount of cross chat will be immense. Consumers will be able to better evaluate the delivered value.
Web advertising will reinforce brand image.	The vast amount of choice will produce more savvy consumers who expect better pricing.
In global commerce, brands will be more recognizable and have a large following.	Vast price differentials will appear where consumers can purchase globally.

signals that consumers will try after they buy (as long as they can return goods), trial is an important risk reducer. Where products can be delivered over the Net, however, the Net can support trial. Many software vendors, including Oracle, now allow users to download trial copies of software, and even full copies for a free (typically) 90-day trial period. Programs self-destruct after a trial period if payment is not made.

Innovations are being instituted that allow for trial. DealerNet, for example, is an on-line service that allows consumers to purchase cars without ever going to a dealership. Buyers collect information on various brands, and then E-mail local participating dealerships with requests to test-drive certain models. Dealerships bring the car to the customer's place of work for a test drive. Digital

Equipment Corporation made a big splash out of allowing buyers to log in to Alpha chip-based computers and run programs on their own data to gauge the computer's effectiveness. Bookstores offer samples of specific offerings, and music stores allow songs to be downloaded and heard.

Purchase

Purchase and the rituals surrounding it are viewed with disdain by many people. Who wants to wait in line? Rude, inept store clerks giving suspect advice and crowded shopping centers and malls are two aspects of buying that many would gladly forgo. Simply allocating the time to go to a store is a hardship for many. For these reasons, direct mail and catalog shopping have become extremely successful.

The Web can ease the purchase ritual. Once a book has been browsed, or a song heard (if trial is even necessary), the shopper fills out the order form (or simply clicks on the desired product) at the provider's site. Customers who have questions can post them to the "inquiries" section of the page and receive responses directly from the provider. Often, the provider is the manufacturer rather than a retailer, and thus offers more expertise in the responses. As emphasized earlier in the book, a revolution in logistics means that shipping the product direct to the customer is a minor concern for both provider and consumer.

After-Purchase Evaluation

Conventional wisdom is that a dissatisfied customer will tell nine close friends about a negative experience. Now, via an on-line discussion group, the person may tell 10,000 or so enthusiasts worldwide. Products may experience drastic and unexpected changes in sales as word of mouth spreads almost instantaneously to worldwide audiences (see the Pentium case in Chapter 6 for an example).

The impact of this on marketing management is profound. An electronic market that provides product and price information enables customers to locate suppliers that better match their needs. Producers may find that arenas of differentiation based on perception or heuristics fade quickly as consumers gain easy access to comparative information. Products are differentiated by quality, price, and details of delivery; and the selling process, as discussed in Chapter 7, becomes an auction. Markets are efficient in the economic sense, because the cost of information, in terms of search and availability, is minimized.

CUSTOMER DECISION SUPPORT

Chapter 2 introduced the notion of customer decision support—providing added value by supporting customer decision making. Table 5–2 applies this notion to our simple model of consumer behavior. At each stage of the process, companies can provide facilities to support decision making.

It is important to note, however, that generic Web facilities, particularly search engines, are available to help the consumer. A company only has control over its own site; a company is reliant on the knowledge infrastructure of the Web to support consumers in finding sites and interacting with other consumers.

For marketing people, the notion of customer decision support signals an important change in their role. The main question now is, "How can I help the customer make an informed decision?" not "How can I persuade the customer to

Table 5–2. Supporting consumer decision making.

Consumer Process	Web Site Support Facilities	General Net Facilities
Need recognition	Advertising on other sites. Physical displays of URLs.	Interaction with other users in discussion and newsgroups.
Information search	Pointers to competitors. Specialized search facilities.	Directories and search engines.
Evaluation	Product information. Trial. FAQs (Frequently Asked Questions). E-mail contact with representatives.	Interaction with other users in discussion and newsgroups.
Purchase	Product ordering.	Financial transactions and security.
After purchase evaluation	Comments by E-mail. Information on return policies.	Discussion and newsgroups.

make the decision I want him to make?" Smart companies have moved marketing efforts toward a focus on customer solutions for a number of years. Like many things discussed in this book, the Net and Web are accelerators and enablers of this change, not necessarily instigators.

LEARNING ABOUT MARKETS

The potential of interactive media to collect data about shopping behaviors at the level of the individual consumer, and to summarize data by individual product, signals an incredible change in marketing management. Purchase diaries and scanner data have given marketers access to actual purchase behavior within a single store, and have allowed them to relate it to richer consumer demographic data. Interactive media, however, allows not only actual purchases to be monitored, but all shopping behavior as well.

Getting the Data

What types of data do Web servers actually collect? When the researchers at the University of Illinois put together the first Web browser (Mosaic) and server, they specified the data that would be passed over when a browser requested a page. Basically, the browser passes over (a) the address of the machine running the browser, (b) the URL of the page presently being viewed, and (c) the URL of the page requested. Since the server knows the time of day, server statistics can thus record the machine wanting the page, where it's getting the link from, the page requested, and date and time. The server does not know the E-mail address of the person using the browser, a common confusion when discussing these matters. Thus if you're on, say, AOL, the server just knows the IP address of AOL. We can work out that you're an AOL customer, but not who you are.

The way to relate visitor information to better demographic data is to get visitors to register. Even if customers are asked to provide minimal information such as age and occupation, this can then be related to all server data generated in following visits. Registration and provision of limited data is now pretty much accepted on the Web. If you allow people to choose a log-in name and number, this can also be used to let them customize what they see at the site. (Conversely, if you force an arbitrary log-in name and password on people, they are unlikely to return.) Storing all this registration data does bring up an interesting cultural problem—many countries, such as the United Kingdom, have data protection laws that limit what can be collected and what the data can be used for. It is unclear whether you are subject to these laws if you collect data from these countries.

What many Web site managers really want to know is, *Where else have you visited?* This is wonderful marketing information—we get to find out not only what pages of ours you've looked at, but what pages of competitors. To allow for this, Netscape has come up with the notion of a "cookie." This is a piece of data placed on your hard disk by a Web site; other sites can request to see all the cookies, thus knowing where you've been (unless you erase the cookie files). This has rightly caused much concern among Web users—no one wants a Web server writing to their hard drive. It's been viewed as an intrusion into personal space, and hence the cookie movement is pretty much crumbs.

Hits, Visitors, and Other Stuff

The major summary piece of data that can emerge from all of this is *hits*. What a "hit" actually is can be defined very differently by different people. We define hit as a request for a page. Hence we can have hits per hour, day, week, or even cosmic cycle. Hits are pretty much rubbish as measures go since they are often inflated by people returning to pages, reloading the page, and so on. A far better measure is *visitor*, which is a request for a page by a machine that has not requested it within the period of measure. Hence, we can talk about number of visitors per hour, day, week, or election campaign, but the number of visitors per week (for example) is not just the sum of the number per day.

Counters that get stuck on pages, such as Web Counter, measure just hits. Through add-on server packages or services like I/PRO, more sophisticated visitor statistics can be generated. Also, paths through pages can be analyzed and summarized.

All this information becomes an evaluation device for measuring the effectiveness of communications and merchandising programs. New performance indicators will be emerging (e.g., the ratio of purchases to visitors, the number of times each day that a particular piece of explanatory information is consulted), that will determine whether or not a product needs different virtual merchandising, needs better explanation, or should be dropped from the virtual shelf. Further, this information can be valuable in rearranging physical sales efforts.

Developing Relationships with Customers

All employees of a provider firm with access to the Net can interact with any customer who posts a question, either through the firm's Web site, or through appropriate discussion groups. This broadens the responsibility for keeping customers well informed and serviced to everyone in the provider organization. When everyone has access to discussion groups and postings to a Web site, employees in every department will hear and consider a response to user questions as soon as they are raised, rather than having them filtered through the sales or customer service organization. The marketing organization may become extremely flat as every member of the firm becomes sensitized directly to the customer's point of view and responds accordingly.

This also poses a major problem: How do you control the amount and quality of interaction when everyone's in marketing? Two options presently exist: Either train everyone to be marketing and sales savvy (expensive and not always possible) or set up a virtual customer support center, similar to existing phone centers (which negates the advantages of having everyone capable of responding). A number of companies have learned the hard way that leaving the timing and quality of electronic response to chance is a disaster. Volvo actually took down their American Web site because they could not adequately deal with the volume of E-mail they received.

Conversely, many network and electronics companies, such as Cisco and 3Com, have successfully combined some generic E-mail support with back up from specialized engineers.

USENET AS VIRTUAL FOCUS GROUPS

To consider the Web to be the sole source of virtual marketing information is a mistake. Discussion groups over on-line services and Usenet newsgroups are where *cross chat* between consumers occurs and can be the source of invaluable marketing feedback. Information that customers exchange with each other is considered the most valuable by the customers themselves, and in the case of newsgroups, we see it in action. It is our opinion that Usenet discussion groups provide a source of primary marketing data that is as yet unmined.

Usenet newsgroups are no longer just occupied by trekkies and Unix buffs. By giving access to newsgroups, and actively pointing people to specific groups, on-line services have changed the demographics of Usenet. As a result, the people posting to and reading newsgroups far more typify customers in the physical marketplace than they did, say, two years ago. And since the number of postings and readers has increased, the amount of valuable information has increased (and also the amount of irrelevant information—but we'll show how to deal with that later).

We can view newsgroups as *virtual focus groups*. A group discusses a topic. When posed a question, a thread of conversation develops. We don't know the exact demographics of the group, and we can't control the discussion or place physical experiences into the group—all things we might do with a physical focus group. But, again, we get contributors from all over the world, discussions take place at an amazingly fast rate, and it isn't costing us anything.

rec.autos.makers.honda

Some newsgroups are dedicated to specific manufacturers, services, and even brands. Nowhere is this more evident than with automobiles. Newsgroups are arranged in strict hierarchies. The root of one set of groups called recreation begins with *rec.*, and this is expanded to *rec.autos*. There is a subgroup called *rec.autos.makers* that has a further subgroup for every manufacturer, including *rec.autos.makers.chrysler* and *rec.autos.makers.honda*. Certain models even have their own groups, for example *rec.autos.makers.mazda.miata* and *rec.autos.makers.ford.explorer*.

If someone's designing, selling, or repairing Hondas, the *rec.autos.makers.honda* newsgroup is a mine of information. Some of the testimonials that people write about their purchases make the response cards used by manufacturers and dealers look like the tepid pieces of market research they are. As a piece of deep feedback on a product, consider the E-mail on page 149.

Subject: Brand New EXa Coupe-Impressions
Date: 2 Jun 1996 21:20:18 GMT

I picked up my '96 green Civic EXa coupe (5 sp, tape, security system) 3 days ago. Today was the first day I got it out on some winding country roads for 2 hours of enjoyment. It rides well: the bumps are felt, but softened. It leans more than my last car ('86 VW Golf) in short curves, does as well or better, in mph, on some local stretches of winding road. It came with Dunlop SP20 FE tires. Are these any good? I like the ventilation system (only needed A/C a third of the time on a day in the 80s), the power mirrors and windows, and the "look" of the car. (Today was also "photo day"). Cresting a hill to find a tractor pulling farm machinery across the road was a good test of the (excellent) brakes. There was a lot of black residue on all 4 wheel covers after just 4 days.

The engine is pretty quiet except for the higher rpms. The shifting seems more positive now than the first day, or it could be I'm more used to it. The key wouldn't turn in the ignition a few times yesterday. Turning the steering wheel didn't help. I had to re-insert the key 2 or 3 times before it would turn. And it really gets hotter inside when parked in the sun than other dark cars I've had—maybe because of the large slanting wind-shield. One thing that could be improved: the lining in the trunk really looks cheap. It isn't flat, and isn't fitted to the trunk walls in the corners towards the front of the car properly. I may just remove it. But, that's minor. Overall, I like the car a lot and look forward to "wringing it out" after the break-in period. Any comments?

The Neon versus the Civic

The preceding example is just one posting. When you follow a thread, the cross chat that emerges is terrific. On June 2, 1996, the following was posted to both *rec.autos .makers.chrysler* and *rec.autos.makers.honda*:

Subject: '96 Dodge Neon vs. '96 Honda Civic
Date: 3 Jun 1996 01:17:36 GMT

I am planning on buying a new car and have narrowed it down to either a '96 neon or a '96 honda civic. I read the edmunds www site and it has good things to say about both the cars. I am worried about possible problems with the neon as told by quite a few people. However, on the other hand, everyone seems to have a honda civic. Further, the neon has more options and power for that price. I will be getting a stick-shift, with A/C, 2 door I would be interested in hearing from people who have had to decide between the two and the reasons behind their final decision.

This was one of 25 posts to *rec.autos.makers.honda* on that day. It's a typical request for information—who out there can help me? (The "edmunds www site" refers to the Edmund's price guide and reviews, which is, by the way, an excellent Web site). Notice the time stamp is GMT (Greenwich mean time). The person posting this is actually in Canada where it's 6:17 A.M. EST. This person certainly got to hear from other people. Here's what happened:

Date: Mon, 03 Jun 1996 03:39:18 GMT

I have recently (February 13, 1996) purchased a new 96 Neon Sport coupe with the DOHC engine and have nothing but good to say about it. The only warranty claim has been for a damaged wheel (I ordered the optional mag wheels). It was taken care of promptly and professionally and I was impressed by this. I would also add that I took my Explorer Post (Part of the Boy Scouts) to tour the Belvedere Assembly Plant (The Only place Neon's are built) and was impressed with the attention to detail and quality. The only caveat with the car is that I would have designed the cruise control and voltage regulation software quite a bit differently. They work, but, I think my 87 Sundance was a little more refined in these two areas.

Date: 2 Jun 1996 22:03:01

The Neon is in all ways a better performer. The Neon is much less expensive. The Neon has more options for the money. The Neon, in coupe form, is damn ugly. Its a great looker in 4 door form, and much more useful. I'd suggest you get the 4 door Highline with ABS, tach, the 150 HP engine, and AC. The Civic is quieter. The Civic has a proven reliability record. The Civic has more aftermarket performance options. The Civic is as plain as plain can be. If you value performance, don't even think about a new Civic. The CRXsi's were great I heard, but the current batch is weak. If you want the perfect car for going 65mph, get the Civic. The Neon is horrible for 65mph. It wants to go 85.

The second posting was from the West Coast of the United States, where it's still June 2. Someone else then puts their vote in for the Neon (with a caveat about style and a sideswipe at Toyota).

Date: Mon, 03 Jun 1996 02:47:34

I likewise just purchased a 96 NEON, and I have to say that I am Very surprised. Originally, I decided on it because of the price (I got the ultimate deal) and now I believe it is the best decision i've made in a while. My brother bought one the same day as I did and he loves his too.

Being frank, don't buy the NEON if you wanna impress the Cobra/Firebird/Z-28 crowd—'cause that just ain't going to happen. But if your looking to get well you're going (and getting great gas mileage on the way) while having fun at an affordable price; then the NEON is for you.

I also looked at many other cars when shopping around. The Cavalier was way too expensive for the base model (which was basically a lawn chair and a steering column); the Toyota people were about $3,000 too proud of their Tercel; and the Honda was just not as much fun as the NEON. The NEON puts something very important back into affordable compacts— personality. Personality was lacking greatly in models like the sundance/shadow, but the NEON is the VW bug of the NINETIES.

Interestingly, the discussion lasted far longer than the decision-making process of this individual. At the end of June, the discussion had turned into Neon vs. Anything, with 4 or 5 postings each day. If you go and look up the newsgroup, it's probably still going on; a good thread never dies. (Incidentally, the guy who posted the original request read every posting, test drove both cars, and then bought a Neon.)

Mining the Newsgroups

For many, the question is how to get at the information in newsgroups. There are now over 30,000 or so Usenet groups worldwide, with many country-specific groups existing locally.

The best way is for someone, or perhaps a couple of people, to be given the task of following a few groups. If you work for Honda of Ford, this is pretty straightforward. There's a newsgroup for your product, plus some vehicle-specific ones, plus a few other more general "lifestyle" newsgroups that you might need to follow. Add in a couple of chat rooms and the like found on on-line services, and the result is that someone, somewhere, needs to monitor about a dozen or so discussions. This probably takes an hour or so a day: the sort of thing that can be assigned to a new MBA recruit.

A problem arises when your product or service doesn't have a one-to-one correspondence with any newsgroup. If you sell life insurance, for example, there's not a life insurance newsgroup where customers just go and love to hang out. To understand your customer, you may be faced with sifting through hundreds of newsgroups that cover topics ranging from death and bereavement to financial planning to smoking. You'll need a tool to help you, and from a research project at Stanford University a tool has emerged called Sift. You specify parameters and areas of interest, it comes up with discussion threads. Search engines like AltaVista now do some of this, but not as well.

USENET FOR ADVERTISERS AND MERE MORTALS

What about getting involved with Usenet? Posting messages and ads? This needs very careful consideration: after the *Green Card* incident briefly discussed in Chapter 2, active use of Usenet to bring people's attention to a product, service, or company has been a minefield. There is still a cultural mentality that marketers should keep out.

Despite this, there are some discussions that corporate types can join in and provide a welcome contribution:

Date: Mon, 03 Jun 1996 19:32:10 GMT

>>The Golf is the quintessential hatchback that all other hatches
>>sold in Europe
>>are invariably compared to. Whenever a new hatch comes into the market from
>>makers like Fiat, the European press proclaims it a "Golf-class hatch . . . ."
>
>Europe, Europe, Europe—what about the rest of the world (no
>offense to our European friends of course)? Who sold more
>hatches from 85–95 in the U.S.? It may very well be Volkswagen,
>but I wouldn't put money on it. Besides, I thought the Renault
>Clio Williams was the quintessential Euro hatch.
>
>Who created the North American hot hatch class? Volkswagen, with that most amazing vehicle, the Rabbit GTI. Whether it is still the class leader is another argument entirely, but it was definitely the class originator.

If someone from VW, or Renault, or Fiat were to jump in with some real figures and their company's perspective, the contribution would likely be welcomed.

Advertising is allowable if it is low key and nonintrusive, just providing a Web address or E-mail where people can get further information. Thus, the following is reasonable:

> Subject:****factory HONDA PARTS wholesale to the public****
> Date: 2 Jun 1996 06:31:39
> Organization: America Online, Inc.
>
> USA Honda dealer
> wholesale prices to the public on factory Honda Parts and
> Accessories
> use our QUICK QUOTE for pricing
> www.stevebailey.com
> or call 1–800–522–3344
> visa mc amx discover personal check

The next example is also acceptable:

> Subject: Wholesale Honda Parts & Accessories
> Date: 4 Jun 1996 15:16:04 -0400
> Organization: America Online, Inc.
>
> We are Brandfon Honda in New Haven Ct and
> we sell original equipment Honda parts & accessories.
> Just E-mail us with the parts your are looking for and we
> will E-mail you right back. Or call us at 1–800–441–4516
> 1–203–772–1493 fax.

The notion of flaming has received lots of attention. Flamers are discussion group contributors who use (or really abuse) the group to vent their anger on some product, service, profession, or person. Some companies now feel that they have to monitor newsgroups to protect their reputation and look for anything that can be classified as slander.

Flaming does occur, and more frequently than it should, but it is amazing how other individuals will step in and defend companies, products, or services, putting flamers in their place.

WEB ADVERTISING

Advertising on the Net looks fundamentally different from conventional advertising. First, the trend toward information reduction required by conventional media is reversed. Conventional advertising is bound by space (for print) and time (for radio and television) constraints, with the goal of creating a memorable perception rather than delivering information. Web pages, however, allow consumers to probe deeper and deeper for more detailed information within a single on-line ad, at their own pace. The implication is that Net advertisers must offer valuable information, allowing browsers to request more as their interest grows. Entire catalogs, detailed product specifications, product performance histories, examples of successful applications, usage demonstrations, and anything else a company wishes to share with customers can be digitized and maintained at a relatively low cost.

Second, the medium is interactive; as such, it allows consumers to actually engage in, and thereby customize, their own advertisements. Information is available whenever browsers need it, is tailored to their interests (since they choose which hypertext links to probe), and allows them to make comments or ask specific questions on-line. This attribute, termed *dynamic addressability*, makes Internet advertising resemble personal sales calls. Specific content can be sent to an individual based on that person's behavior. This can be seen with search engines, which serve their banner ads based on what a person is searching for.

What causes a disinterested surfer to search out the ad in the first place? One alternative is to make users aware of the page by posting its address in a relevant discussion group. The audience will be the narrowly focused target that already shares an interest in the advertiser's product. While this is efficient, it still does not gain the attention of those who do not yet share the interest. Second, URLs can be present in all other forms of advertising: TV and print ads, collateral material, packaging, even business cards and the company letterhead. It's rare to see a physical ad that doesn't include a URL for the company or product.

A final, broader alternative, is to arrange for paid sponsorships of news content sites, or hypertext links to the ad from other pages that are accessed heavily. The graphic that is used as the link must be unusual enough to attract attention and cause the browser to want to go to the ad. It is this arena that poses the greatest challenge to advertisers on the Internet: how to entice disinterested people to browse the Web page. Promises of free samples, random drawings, and other enticements have all been used; Cathay Pacific's offer of a drawing for 2 million free air miles springs to mind.

Tracking Ad Effectiveness

Tracking the effectiveness of advertising has always been a challenge. Now data are available that will enable this. The number of visitors to a Web site, the number of repeat accesses from a single visitor over time, the number of visitors that click from a page into subpages and the amount of time spent by a visitor to any single subpage are data that can all be captured (see the section Hits, Visitors, and Other Stuff, earlier in this chapter). These statistics can be used to measure reach, frequency, and depth of interest. It will also be easier to distinguish the effectiveness of the *ad* from the effectiveness of the *product* being offered. The number of visitors to fun or interesting sites means effective advertising, independent of whether or not a product is purchased.

Commissions based on media billings may be a thing of the past. Procter & Gamble has insisted that they pay advertising rates on the Web based on click through: the amount of traffic that the carriers of advertising send to their site. Since, as discussed earlier, they know where the traffic to their sites is coming from, it can be calculated without the help of the advertisement carrier. This is likely to become the dominant model.

How much money can someone make carrying advertising? Is there enough advertising out there to be able to make money just from carrying ads? Very rough estimates of first quarter 1996 figures for advertising revenue and expenditure in the United States, for well-known sites and relatively big spenders, are:

Revenues: InfoSeek and Lycos: $3 Million
Yahoo! and Netscape: $2 Million
c|net, ZD Net, ESPNET, Sport Zone,
Pathfinder, and WebCrawler: $1 Million

Expenditure: IBM: $1.5 Million
Microsoft: $1Million
Netscape, c|net, AT&T, Nynex, MCI,
Internet Shopping Network, Saturn, and
Excite: $400,000

This looks like a bit of a closed shop. Both Netscape and c|net make money from advertising and spend it. While these numbers are growing, total Web advertising would only buy a couple of broadcast Superbowl ads.

Cultural Problems

Considered as a cultural phenomenon, the Web presents a particular challenge to advertisers. Accustomed to developing campaigns and advertising material at a local level, ads placed on the Web are now viewable worldwide. For companies like British Airways, with a global brand, this is an opportunity to present a global image. To other companies that have different products in different markets, it is often not clear whether virtual advertising should be local (as with physical campaigns) or global. Due to the advanced nature of Internet working in the United States, and the importance of the U.S. market, companies like Toyota and Canon have created Web sites that are firmly aimed at an American audience. Conversely, some global companies such as General Motors have created a global "super site," and then allowed divisions in different countries to control their image and Web advertising. The Vauxhall division in the United Kingdom, for example, has been reasonably aggressive in pushing its singular Web presence.

The interactive nature of the advertising, where digital samples can be attached to ads, is similarly a challenge. In late 1995, *Redbook* placed the click-through banner advertisement with the search engine Lycos. (For those outside the United States, *Redbook* is a women's magazine positioned somewhere between *Cosmopolitan* and *Good Housekeeping*, i.e., sex, clothes, *and* recipes.) While reasonably racy, we are used to ads such as this in print mediums. However, the click-through sent the visitor to the actual article from *Redbook*.

Jim Hopper, a PhD student in clinical psychology at Boston University, was incensed by this ad. His research had shown that victims of child abuse had great difficulty confronting sexually explicit material, and the last thing they needed was banner ads on the Web enticing them to read such material. Moreover, a major user of search engines comprises schoolchildren trying to find information to help with homework. He complained to both the CEO of Lycos and HomeArts, the publisher of *Redbook* and part of the Hearst Corporation. To their credit, HomeArts saw that the ad and the ability to click through were providing material to those who would not normally, and perhaps should not, see it. They withdrew the ad. Jim received the following E-mail:

Dear Mr. Hopper,

I have received your inquiry about our HomeArts advertisement on the Lycos search engine. HomeArts does not intend to promote mature themes to young audiences. We began our advertising with Lycos under the assumption that this service would provide a good way to reach a wide base of Internet users. We were not aware that this "wide-base" could include younger people on school assignments. When we realized the mismatch, we withdrew the advertisement and replaced it with more family-oriented promotions. Such considerations will be a key component of our advertising decisions in the future.

I'd also offer that the Internet is a new medium and we, like everyone else, are still learning how to properly target our messages. Your input helped this process along. Thank you.

Sincerely,

Brian J. Sroub
VP, Marketing

The key word here is "target." The short history of the Net and Web shows that, from the *Green Card* lawyers to this incident, everyone who doesn't target their audience trips up. Advertising on the Web should be targeted, not blanket. In fact, now that search engines place banner advertising on the pages sent to browsers based on search words and results, the *Redbook* problem is less likely to happen.

This instance probably did not make Jim Hopper very popular among some groups. The cherished right of free speech on the Web extends, in the minds of many, to protect anything that is legal, including the *Redbook* ad. Our take on this is that advertisers would do well to consider who will get to see their message, and how they are going to communicate with prospective customers without intruding into the lives of others.

Cross-Medium Campaigns

If blanket advertising on the Web with banners that might offend some is dumb, then not linking virtual advertising into physical campaigns is even dumber. Companies such as AT&T, NYNEX, and British Airways have done an excellent job of balancing the look, feel, and message of their Web sites with their physical advertising. AT&T creates subpages directly linked to campaigns. For example, perspective purchasers of their TrueVoice service were, for the duration of the advertising campaign, able to visit http://www.att.com/truevoice.

It is thus amazing how many Web sites *do not* convey an image anywhere near that in their physical advertising. For some time, a poor site that was identified as poor in two separate talks by advertising people was the one for Frito-Lay. The (soon-to-be former) PepsiCo division has extensive TV and print advertising aimed at young people. The Web site made two incredible marketing errors. First, its front page used colors and graphics that were entirely different from the physical advertising. Second, the material that followed was aimed at adults. Visitors could read how FritoLay was a "leader in logistics" with a long and proud company history. Boring, boring, boring—and we're not even teenagers!

If a company does want to convey multiple images to multiple audiences, then it should create multiple sites.

Procter & Gamble does not have a company home page—
they create pages for each product line. Similarly, in some
companies separate sites for shareholders, corporate investors, and employees may be appropriate. A company
does not have one physical contact point, and there is even
less reason for this to be the case in the virtual world.

BEST PRACTICE

Best practice in virtual marketing is a tough one—poor
products and services can be the subject of great marketing, and vice versa. Rather than specific sites, we suggest
here two companies that are leaders among the new electronic advertising agencies, plus one company that's a bit
of a surprise and not immediately obvious, and one that's
a shoo-in when it comes to marketing.

- *Free Range Media.* Based in Seattle and now employ-
  ing over 50 people, Free Range develops Web sites for
  various corporations. Their sites have ranged from
  pure advertising and entertainment (the NFL and a
  CompuServe site for Baywatch fans), delivery of content, a physical medium (CBS's Eye on the News and
  the excellent *Christian Science Monitor* site) to hotel
  booking and value-added services (Macmillan Publishing, Prentice Hall Direct, Westin Hotels). With a
  staff that includes professional technical writers and
  graphics people, they always seem to be able to create
  sites that are appropriate for the market, fit the image
  of the client company, and look great.
- *Organic.* Based in San Francisco, Organic is similar to
  Free Range, but perhaps places a greater emphasis on
  advertising (Free Range also manages servers and the

technical side of operations). Clients include many brand-name companies, and companies that own well-known brands: McDonald's, Colgate-Palmolive, Levi Strauss, Nike, and the Saturn division of General Motors, to name a few.

The sites created by both Free Range and Organic are exemplars of the state of practice. They employ animation using Shockwave (a way of doing sophisticated gamelike animation over the Web), the Java programming language as appropriate, and often wonderful graphics.

- *Netscape.* Why are they in this list? Well, for a start, the way they've offered the Navigator browser for trial and beta test through downloading off the Web has revolutionized the way software is sold. Other providers of client software have had to follow this lead. But they're also here because their use of Usenet newsgroups is exemplary. To grow, they needed to carry the technical Net population with them, and have provided excellent support through E-mail. They have taken the small company mentality that everyone's in marketing and kept that as they've grown. Further, they've used the *comp.infosys.www* newsgroups to great effect. Imagine being in the middle of a technical discussion on, say, Web protocols, when Marc Andresson (cofounder, now VP for Technology) posts a message explaining Netscape's position. You can bet that everyone reads it, and feels good that they've had some communication with a Web star.

- *Procter & Gamble.* In the same way that anyone in retail should keep an eye on Wal-Mart, anyone in marketing should keep an eye on P&G. They were slow to enter the Web, but have done so on their terms. They first gained attention when they registered domain names for over 100 of their brand names. As mentioned earlier, they have developed sites for products (not the company) and negotiated payment based on click-through.

6

The Medium Is the Message

In June 1994, Dr. Thomas R. Nicely, a professor of mathematics at Lynchburg College, Virginia, noticed a small difference in two sets of numbers that had been generated on two different PCs. After various tests with different software and hardware, he became convinced that the error was caused by Intel's Pentium processor. On Monday, October 24, he contacted Intel technical support to report the problem. The person he contacted duplicated the error and confirmed it. But Intel did not acknowledge the error, and did not take the matter any further.

What then happened until the end of 1994 is now well known. The error in the Pentium chip led to very bad press coverage for Intel, a spate of Pentium jokes and, more seriously, a drop in Intel's stock price. What makes the episode interesting is that the Net was the medium by which word about the problem spread. It is a documented case of one person telling thousands of others about his or her evaluation of a product (discussed in Chapter 5), but in this case the thousands ended up being millions of people. What is most interesting about this scenario is Intel's response. The company could have responded on the Net and reached those interested millions and thereby minimized the damage, but did not. In hindsight, Intel's reaction was bad for business—really bad.

PUBLIC RELATIONS, INTEL, AND THE PENTIUM FLAW

Public relations, press releases, interaction with analysts and important industry observers—in many businesses, this is the work that keeps the fax machine busy. Draft a press release, put it on the fax machine, hit a button that pulls up a few dozen predefined phone numbers, and you're in business. The trouble is, in many industries the Net is now the preferred medium of customers, analysts, and journalists. They seem to have one up on some PR staffs and firms, whose idea of using the Net is to blanket E-mail the same old boring press release they're still faxing. The catchy title, the quote from the boss, the phone number for further information. The world has changed, and the savvy have realized this. But back to Intel.

Following his first interaction with Intel, Dr. Nicely sent an E-mail message to a few people explaining the flaw in the Pentium chip. He wanted to find out if anyone else had discovered it. By Tuesday, November 1, Nicely's E-mail had been forwarded to Richard Smith, president of Phar Lap Software in Cambridge, Massachusetts. Phar

Lap's programmers tested and confirmed the division error. Realizing the significance of the flaw, and wanting to keep his customers abreast of the problem, Smith forwarded Nicely's E-mail to important Phar Lap customers and other companies, including Microsoft, Borland, Metaware, and Watcom.

On Wednesday, November 2, news about the flaw reached Terje Mathisen of Norsk Hydro in Norway. Mathisen confirmed the flaw, and then posted a message to the Usenet newsgroup *comp.sys.intel* titled "Glaring FDIV Bug in Pentium!" (FDIV is the instruction that does floating point division, i.e., divides two real numbers together.)

On Monday, November 7, an article ran in *Electronic Engineering Times* in which Intel said it had corrected the flaw in subsequent runs of the chip. On Tuesday, November 22, Tim Coe of Vitesse Semiconductors and Mike Carleton of the University of Southern California announced on the Net that they had reverse-engineered the way the Pentium chip handles division and created a model that predicts when the chip is wrong. Intel's stock droped 1⅜ points. CNN's *Moneyline* program picked up the story and ran it.

On Thursday, November 24 (Thanksgiving holiday in the United States), the *New York Times* ran a story entitled "Circuit flaw causes Pentium chip to miscalculate, Intel admits." A similar story by the Associated Press was printed by more than 200 newspapers and ran on radio and television. Intel offered to replace chips, but only if it could be proved that you have an application in which the error would occur.

Finally, on Sunday, November 27, a posting appeared on the newsgroup *comp.sys.intel* from Intel's president, Dr. Andrew Grove (but bearing someone else's return address). The posting was part an admission of guilt and part surprise about the furor. If anything, the posting made things worse. On Monday, December 12, IBM issued a press release stating that they would halt the shipment of

Pentium-based PCs. On Friday, December 16, Intel stock closed at $59.50, down $3.25 for the week. On Tuesday, December 20, Intel finally apologized and announced that it would replace all flawed Pentiums on request. It set aside a reserve of $420 million to cover costs.

What went wrong? Intel, perhaps inadvertently, broke two of the golden rules of maintaining customer and public relations on the Net. First, they did not contribute to (and maybe did not even monitor) Net-based discussion. If they had quickly responded to the postings that appeared in *comp.sys.intel*, much of the bad press might have been avoided. Second, they broke the "crossover" rule: they only posted a response to *comp.sys.intel* after the story had moved to the mainstream physical media. Once CNN had got hold of the story, a full physical response was required.

Intel is a leader in the semiconductor business; its microprocessors power most of the world's PCs. By early 1995, its stock had recovered and surpassed its level prior to the Pentium episode, and its entire 1995 production of Pentium chips were under order. But Intel learned its lesson; it now posts all known flaws to newsgroups, and maintains a staff that follows newsgroups and discussion groups on on–line services. When a flaw was discovered in the Pentium Pro, Intel announced it before it was discovered by customers, both on the Net and through physical press releases.

THE GOLDEN RULES

The present problem with PR and the Net is a wonderful example of a culture clash. Proper and successful use of the Net requires that the goals of PR (presenting a message that puts forward a point of view, getting new information out, engaging media journalists, and getting attention, etc.) have to be aligned with the cultural requirements of the Net, as discussed in Chapter 2. Our experience suggests that many people in PR have absolutely no idea how to do this. Their use of the Net ranges somewhere between incompetent and unethical. They E-mail people they've never talked to before (either virtually or physically), they post intrusive messages on Usenet, they join mailing lists and then preach rather than contribute. Using the Net as a PR vehicle requires immersion in its culture.

If certain rules are followed, the Net can become a wonderful and effective PR tool for business. We have mentioned two golden rules; here we expand these to five guidelines. They are summarized in Table 6–1.

Rule 1. Listen

Make sure that you follow, and perhaps contribute to, every Usenet newsgroup and on-line service discussion group that is germane to your business. You cannot make

Table 6–1. Good PR on the Net.

PR Rule	Explanation
Listen	Follow postings in newsgroups and discussion groups. Know what's going on.
Use the right medium	Don't counteract physical reporting in virtual mediums, and vice versa.
Use the Net to target	Target your message; don't carpet bomb.
Cut up your material	Send E-mail that invites people to get further information. Entice, don't intrude.
Respond	When someone replies, quickly and diligently respond. Engage people; don't ignore.

pronouncements if you have not listened. You will learn what people are concerned about, what gets them angry, and what makes them impressed with a product or brand.

Chapter 5 included an extensive example taken from the perspective of marketing. From a PR viewpoint, much of what was said there is applicable. If you read extensive amounts of commentary from unhappy customers, chances are you'll know better how to deal with them.

Rule 2. Use the Right Medium

When companies want to make us aware of their Web site, they rightly E-mail us. But occasionally, a PR firm will phone us and ask for a fax number where they can send a press release. If you can't provide material in the medium that people live in, forget it. In some instances, you may have to do some legwork to get the medium right. It can be worthwhile finding out how someone wants to receive information prior to actually shipping it out.

Similarly, if you need to put out flames (like Intel) choose the medium where the fire is biggest. Don't expect to be able to counter reporting by CNN with postings to newsgroups or a few E-mails. Conversely, don't call a press conference to denounce a competitor that is libeling you on the Net.

Rule 3. Use the Net to Target Your Customers

If one theme runs throughout this book, it's that the Net is an opportunity to carefully interact with individuals. This is nowhere more apparent then in PR. The advantage of the Net is not that you can reach more people, but that you can tailor your message. So don't E-mail 1,000 people rather than fax 100. Use the medium to tailor what you send those 100 people and get their attention.

The effort required to do this well is considerable. But this has to be balanced against the impact. Carpet bombing on the Net is, in our experience, at best irrelevant and at worst detrimental. A PR firm, hired to announce a new Web service, E-mailed a press release to individuals and posted it to newsgroups. The tone of the release was, to say the least, aggressive. A senior executive in a position to do business with the service was so incensed at receiving multiple copies of the message and coming across it in newsgroups that he complained to the company that was offering the service. The worst PR is PR that backfires.

Rule 4. Cut Up Your Material

Another common mistake is to use E-mail to deliver huge messages. The typical one-page press release is too big for E-mail: Most people get to see the header and title in their E-mail reader, and if that doesn't grab them they don't read on. An E-mail message should be 5 to 6 lines *at most.*

So how do you get deeper material to people? You put it on a Web page, and provide a URL in your first message. Or you tell people that if they reply to the message, they will receive the full release. Or you send it as an E-mail attachment that they can print out or view on a screen.

Rule 5. Respond

Have you ever responded to a press release E-mail? Ever send back a message along the lines of "This looks interesting, can you give me more details?" What commonly happens when you do is— nothing. Traditional PR is often perceived as a one-way method of distributing a message. When do you ever contact PR people, rather than have them contact you? But the Net is different—the effort and cost of replying is so low that people will do so. This opportunity to engage people in a dialogue is surely the goal of good PR.

We're not PR people, but we have worked with them in the past. We know good PR when we see it. On the Net, we've seen two consistently good users of virtual PR. One involves (which we'll come to in a moment) an individual who has started a company with the objective of using the Net to target messages. The other is a company that has used the Web to fight some battles.

WAL-MART: US AGAINST THE WAL

Wal-Mart seems to keep cropping up in this book. Maybe they're just so big and visible we notice what they do, but then again, maybe they just know what they're doing.

U.S. readers will be aware of the problems that Wal-Mart has encountered when trying to build one of its monolithic stores in certain local communities. Wal-Mart's consumer goods prices are so competitive that often they can sell products for less than a small retailer can obtain them from a distributor. A Wal-Mart development application is thus sometimes perceived as a forecast of the demise of locally owned retailing. Hence local retailers and residents have organized and fought the possible construction of Wal-Mart stores, fearful that downtown shopping (and even downtown itself) will be destroyed once the store is opened. Planning boards have found themselves in a difficult position: Granting a permit to Wal-Mart, may very well adversely affect local retailing, but not granting a permit means forgoing the jobs created by the store, the tax revenue, and the cheap shopping afforded to residents. And if Wal-Mart then just constructs a store in the town next door, retailing may still be lost anyway as well as the tax revenue. There is never an easy or obvious solution.

What's this got to do with the Net and Web? For activists, the Net is a very attractive medium. Most local activists have two big problems: they have no money, and they need to get their message beyond the immediate community. Putting up a Web site, using E-mail, and posting to newsgroups all provide a low-cost method of distributing a message. Activists no longer need to find the cash for a printed pamphlet or phone calls, they just need an account with an ISP. They can E-mail anywhere in the world, and their Web site is globally accessible.

Figure 6–1 shows the front-page of the Web site constructed by the Peninsula Neighborhood Association (PNA) in Gig Harbor, Washington, to aid their campaign to stop Wal-Mart building in Gig Harbor. This is typical of the battles that have ensued as Wal-Mart (based in Arkansas) has moved northward in its march through American retailing. Was the Web site of any use? It certainly got some attention: The site was featured by *USA Today* on their Web site, and the local press made something of it. At the time of writing, the battle of Gig Harbor is at a stalemate, with Wal-Mart having been denied development permission first time round but still retaining an option on the site they wish to develop.

In response to activists who have put up Web sites criticizing Wal-Mart or fighting possible Wal-Mart developments, Wal-Mart has in some cases created Web pages

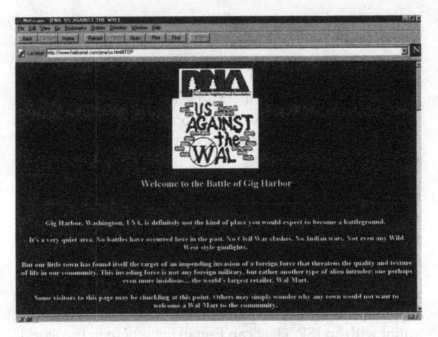

Figure 6-1. Fighting Wal-Mart On-Line.

answering criticisms. One such example was a proposed development in Virginia where residents claimed the development would destroy the land where George Washington cut down his "cherry tree." Wal-Mart countered with Web-based maps and graphics that begged to differ.

BEST PRACTICE

Writers have leaped on instances of bad practice in virtual PR, including the Intel story and Volvo's decision to stop answering E-mail from customers. It's time to look at a positive example.

Eric Ward's NetPost is an effective virtual PR firm. Based in Tennessee, Eric formed NetPost to help companies with Web sites get listed on other Web sites and in directories. Although his business focuses on PR for Web-based companies, his approach is (we believe) widely applicable.

There are services that will automatically send a URL and a description of a Web site to various directories and listings. While this is a cheap way to get yourself listed, you have no control over where the directory lists you and when they do it. Moreover, getting into a directory is not going to get attention for your site. As a new virtual business, pretty much all you've done is told the phone company you're open for trade.

Conversely, a mention in a magazine or a review *will* get you attention. Thus new Web-based businesses, and physical businesses moving to the Web, have a need for PR. They need to get into the reviews, the on-line magazines, the selective lists that point surfers to what's really cool or useful. This is where Eric Ward comes in. Eric's approach has been to strike up relationships with the people who run selective lists and electronic magazines as well as physical magazines and newspapers that focus on the Net, and then send them press releases when appropriate. He calls this part of his service NetWIRE. The people he deals with are also global—he can get a Web site that has been developed in one country reviewed in another.

Eric doesn't just blast out copies of the release; he releases them with a personalized E-mail aimed at both getting attention and continuing the relationship he has developed. For example, this is what we got from him preceding a press release concerning the new Web site for The Flying Noodle pasta store:

Date: Tue, 1 Oct 1996
From: "Eric Ward's NetWIRE" <netpost@netpost.com>
To: Bob O'Keefe <bob@owi.com>
Subject: NetWIRE\SiteNews\Flying Noodle Pasta Lovers Site Opens Subscriptions

Bob,

Below is an announcement from one of the Net's better known food marketers, The Flying Noodle. I typically don't send you releases when they are about sales oriented sites or promotions, because there are a few million of them <g>, but of special note is that The Flying Noodle does over 30% of its revenues from on-line sales alone, which has some positive implications from a targeted on-line commerce perspective. They only market gourmet pasta and sauces, and also have a newsletter with over 300 subscribers, and (I'm not making this up) an Internet Pasta Club. Not bad for such a defined niche. Since launching last year, the site has been covered in several publications, including USA Today and NetGuide.

Company owner Raymond K Lemire (The Big Parmesan) believes that a goal of 50% of revenues from on-line sales is within reach by next summer.

Please feel free to review this site news release, and if you like, to consider as an Interesting Business Site on the Web.

Thanks for your time, —Eric, the NetWIRE

This personalized approach is time consuming. But in our view, it works. We read what he sends us since (a) the prefacing paragraph is always worth a read, and (b) we trust him to send us only releases that we're interested in (as mentioned—we get more sales-oriented stuff than we can look at). Thus, he actually provides a service for us by targeting us when appropriate. At time of writing, I have a mailbox that contains 26 press releases from people wanting us to look at Web sites. Three of these are from Eric. I am unlikely to look at all 26, but guess which ones I'm going to look at first?

We asked Eric to describe his business indicating what differentiates him from traditional PR firms. His reply includes insights into directories, relationships on the Web, and the global nature of the medium:

I've had several success stories, some for big sites, and some for very small sites, which actually is the most rewarding part. I love it when a small business, like the Flying Noodle, or Scambusters, has a big response to my campaigns. Many of my small clients have appeared on TV and radio shows that are popping up about the Web and Internet, as a result of my campaigns. I have dug deep to locate and build relationships with the producers of these shows, and I bet the traditional PR firms don't even know they exist.

The part of the business that people don't get is that it is never going to be possible to find a master-list of Internet/Web media folks that need to be contacted. Whether it is print magazines about the Internet, which are popping up like wild all over the world, or Web-zines that appear out of nowhere and suddenly have hundreds of thousands of readers. Or the emerging area of news filtering services that present sites to people based on predefined criteria. Or the new TV and radio shows about the Web, or Real Audio shows. The list of media people and places that are interested in learning about Web sites and Web events is staggering, and nobody will ever have all of them. I have nearly five hundred contacts in 10 countries, who do nothing but cover the Web, Web sites, and Web events for their particular publication or program or zine or you name it. By next week there will be many more.

Still don't think that the Net is a medium for *targeting* a message that is well thought out and polite? Eric's views on this are even more extreme than ours:

> Just because a group or niche exists that appears to be perfect for your message does not mean you should send it to them. Mailing lists and Usenet groups were not created for marketers, so respect these forums by actively participating, not selling.

E-mail press-releases are absolutely an art form, and media E-mail addresses are a personal canvas you must brush very carefully on. They are far more sensitive than fax machines. A hit-or-miss approach via E-mail to media folks is rude, inconsiderate, and won't work.

7

Strategies for Net Retailing

$\mathbb{R}$etailing in the United States is unrecognizable from the retail landscape of 20 years ago. Many locally owned stores in downtown areas have closed since the late 1970s as retail moved to out-of-town shopping malls. Large discount retail operations such as Wal-Mart and Kmart have removed shopping even further from the community. At the same time, catalog shopping has boomed following direct marketing leaders such as L.L. Bean and Spiegel, and home shopping through television has emerged as a major retail force. Home shopping through cable channels now generates about $2.5 billion in the United States (although this is less than 0.2% of total U.S. retailing). It was once thought that shoppers would not buy items that they could not feel and try, but catalog and home shopping sales have proved this wrong. If shoppers can return items without question, then they will purchase prior to trial.

It's no surprise that the rapid changes in retailing have been enabled by technology. The use of advanced logistics and communications by mass merchants such as Wal-Mart, the analysis of vast quantities of scannable data, and the provision of store-specific credit cards are just three examples. Information to aid in inventory planning and control now flows through the value chain at a speed that allows retailers to act quickly on changing customer requirements; stores that would stock for a season (e.g., winter clothing) and then analyze the success of sales after the season can now replenish and alter items in response to consumers on a, say, weekly basis. This innovation has impacted the entire manufacturing industry, which is made up of layers and layers of suppliers and vendors.

The newest part of this ever changing IT-enabled retail landscape is on-line marketing, particularly retail through the Web. In Chapter 5 we looked at marketing and advertising, and the impact of the Net and Web on marketing initiatives. Now, we'll consider just plain retailing—what companies are doing to set up shop.

ON-LINE CONTENT

Despite the youth of the Web as a retail medium, some common patterns and paradigms are emerging. A virtual shop really has two dimensions: *content* (the way in which the product or service is presented) and *transactions* (the type and extent of on-line processing that is performed). The basic approaches to content are shown in Table 7–1. Content is far better understood than transactions; we'll

Table 7–1. On-line content paradigms and typical transaction processing.

Content Paradigm	Basic Approach	Typical Associated Transaction Processing
Electronic billboard	Web page as advertisement	None
Electronic brochure	Deeper information over multiple pages	On-line order form
Virtual catalog	Multiple products arranged as a catalog	On-line order form, shopping basket, payment
Inverted catalog	Useful information with sales offered below information provision	On-line order form, shopping basket, payment
Virtual mall	Mall provides a home for multiple shops	On-line order form, shopping basket, payment
Electronic bazaar	Members can buy/sell/trade information, service and products	Reconciliation amongst members
Electronic auctions	Auction product, service or contract to highest bidder	Acceptance of bids, notification of awards
Electronic brokerage	Web site acts as a broker for consumer requirements	Electronic interaction with market suppliers, delivery, payment

identify the key approaches to content and then return to consider transactions. Managing transactions requires rather more sophistication in the use of technology and its integration with business processes.

At the simplest level, a Web page can be considered as an *electronic billboard* or *electronic advertisement* that provides information on products or services plus contact information for interested consumers. The majority of present commercial Web pages probably fall into this category. Such advertising may be useful in niche markets where just getting the word out is important, or where a company wants to provide deeper information than can be contained in regular advertising. Consumers can also be enticed to order physical catalogs or samples. Where product information is provided at a level deeper than would normally be seen in a physical advertisement, the term *virtual brochure* is sometimes used.

Beer and alcoholic beverage companies, such as the Canadian brewing company Molson, have taken this type of advertising to a new level. Molson's site provides entertainment, including the ability to take part in discussions and E-mail other visitors. Users have to register for the site, providing Molson with demographic data. Web sites for specific brand names are also common, including Zima and Captain Morgan Rum. Because "hard liquor" companies in the United States have a self-imposed moratorium on TV advertising, it is no wonder that they have jumped at the chance to use the Web. What entices people to visit these sites other than entertainment, competitions (such as random drawings), and printable coupons?

Virtual catalogs typically couple information on products or services with facilities for ordering, much like a physical catalog. Advantages that accrue to virtual catalogs compared with their physical counterparts include:

1. The catalog can be linked to inventory data, so that the user can see if an item is immediately available or not (as in the Nine Lives example in Chapter 1).

2. A company can quickly add new items to a catalog, without waiting for the next catalog printing.

3. Items that sell out and cannot be replenished can be removed immediately. This is especially useful for items whose supply is naturally limited, such as antiques or rare automobile parts.

4. Items that are moving slowly through other channels can be offered at special discounts.

5. The consumer can be provided with search facilities to quickly locate items. This is especially useful for producing catalogs with thousands of items, such as compact discs or books.

Numerous examples of virtual catalogs now exist. Kaleidospace sells various types of art; Speak-to-Me is a catalog of toys that make noise, and samples of the noises can be downloaded. CDNow! has become a major retailer of compact discs; EMusic is a similar service that provides excellent search facilities. Amazon.com is a virtual bookstore that aims to provide a huge selection of books, providing the visitor with both powerful search facilities, and (if you can't be bothered to stick around and look for stuff) agent technology that will E-mail you when a book is published and available. Figure 7–1 shows the storefront of Software.Net, one of the more exciting retailers, discussed in detail later.

Catalog leaders such as L.L. Bean, Spiegel, and the seed company Burpee have been quick to seize the possibilities of virtual catalogs. They offer special deals on slow-moving items, and can also test sales of items prior to putting them into their physical catalog.

In a typical catalog, the product is put "up front," and associated information about the product line or its derivation may be available elsewhere or very limited. The Web, with its cultural history of providing free information,

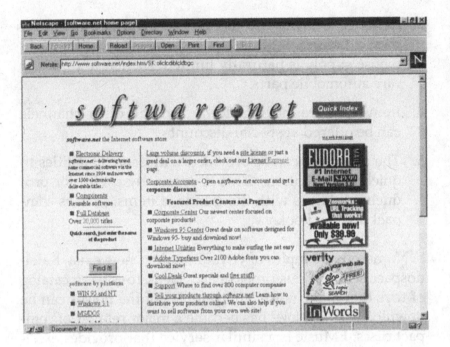

Figure 7–1. One of the New Retailers—Software.Net.

allows catalog marketers to invert their catalog. In this paradigm, relevant free information and services are provided to the consumer who visits the page, and it is not always apparent that a product is being sold. The user, searching for more information, may then be told about a book, compact disc, or financial product.

Perhaps the earliest and best example of this approach is The Real Beer Page, a start-up company that provides its users with information on microbreweries and their locations. Valuable as an information source in its own right (users can get lists of breweries in geographic areas that they define), the page also markets beer-associated memorabilia such as T-shirts. BMG's Classical Music differentiates itself from other compact disc catalogs by providing considerable information on composers, instruments, and periods. A user can find information about, say, Mozart symphonies, and then be offered a selection of available discs. The Parent's Place (like the Real Beer Page, a small start-up company) provides information and links to parenting resources while selling books, cassettes, and other materials on parenting.

For a business that does not want to manage a Web server itself, locating in a *virtual mall* is a viable option. Like a physical mall, the mall owner rents out space, draws traffic to the mall through advertising, and sometimes manages orders and order transactions. There are now over 400 established virtual malls worldwide; interesting examples include the Internet Shopping Network (perhaps the first serious mall on the Web), the London Mall, the EUROMALL, and Downtown Anywhere (an early entrant). Like physical malls, a shop can actually locate in more than one mall.

Virtual malls, unconstrained by physical space, are becoming multifaceted venues that provide information and entertainment as well as shopping. *Virtual bazaars* take this a stage further—they allow members to sell, buy, or trade with each other. Hence the bazaar becomes a focal point for trade and interaction between members. Industry.Net is a very successful example focusing on business-to-business relationships between small and medium-size manufacturers. In fact, Industry.Net can be viewed as a continual virtual trade show, and the trade show is probably the twentieth-century version of the bazaar. Members can buy or sell, for example, a used injection molding machine. (AT&T has signed a deal with Industry.Net to take this model into other areas.)

Perhaps not a content paradigm in its own right—more a method that can be added to catalogs and other paradigms—*virtual auctions* model physical auctions, except that bidders are stationed on the Web, and the auction can proceed for a matter of days or weeks. Like physical auctions, items can have reserved pricing, and the auction method can be either the common highest bidder approach, Dutch (where the offerer reduces the price in increments and the first bidder gets the item), Japanese (where the offerer, selling multiple lots, increases the price so that early bidders get a better deal), or some more complex approach. Bids can be sealed (a bidder only sees his or her bid amount), or open, or bidders can be given some feedback on the range and type of bids.

The airline Cathay Pacific has developed what it calls *Cyber Traveller*, where customers can bid for seats on certain flights. The auction is fairly well bounded. Table 7–2 shows a status report from an auction. An interesting aspect of

Table 7–2. Bids in a Cathay Pacific seat auction for return flights from Los Angeles to Hong Kong (Travel dates were restricted.) CyberTraveler Auction #3 status report generated on 08/10/96 14:05 PDT.

Item	Economy	Business	First	Total
Seats up for bid	313	56	18	387
Minimum bid permitted	$ 300	$ 600	$1,200	n/a
Maximum bid permitted	$1,100	$3,000	$6,000	n/a
Total bids received	9,934	3,589	1,237	14,760
Amount of highest bid	$1,100	2,700	$4,700	n/a
Number of bids at this highest bid	14	1	2	n/a
Number of bids at the minimum bid level	585	376	158	n/a
Average bid amount	$ 482	$ 978	$1,825	n/a
Number of bids above avg	4,472	1,535	444	n/a

electronic auctions is that whereas in a simple physical open auction each bid in time is known to all, the cyber auctioneer can be fairly sophisticated in deciding what information to impart. The data in Table 7–2 are fairly generous—someone wanting to place a realistic bid should be able to make a good guess at the limits of possible acceptance. American Express has experimented with Japanese auctions, enticing customers to buy tickets early before prices increase.

In an electronic brokerage, the retail outlet acts as an intermediatory between consumers and suppliers. InsWeb is one such example—visitors can specify the necessary information for an insurance quote, and InsWeb then obtains the quotes. Present operations are limited to automobiles and the state of Utah, but this is a test prior to establishing facilities for more states and also niche insurance markets (such as boats). Insurance Matters is a similar service based in the United Kingdom, delivering quotes electronically.

Electronic Markets

So, what is an electronic market? This phrase has been attached by numerous people to numerous things, ranging from EDI (which is definitely not any sort of market) to electronic brokers. The theory suggests that since transaction costs can be lowered and market information occurs in real time, the electronic market works more efficiently.

Although we have also thrown this phrase around (it's great for grabbing the attention of senior management), electronic markets that presently exist tend to be well defined, such as electronic versions of futures trading where software may make many of the buy-and-sell decisions. A market should have some role in determining the price of a good, so simply offering goods for sale at a fixed price is marketing, not creating a market. We find it makes more sense to talk of an *electronic marketplace,* where customers and providers meet, exchanging goods and services. As discussed in Chapter 2, the Web can be perceived as one big marketplace.

A market has to have some underlying framework for operation—and electronic bazaars, auctions, and brokerage are such approaches. Others will no doubt emerge. It is not unusual for early uses of technology to be based on physical counterparts, as are these, and only later for these ideas to be stretched. Think what happened when word processors replaced typewriters ("no more correction fluid!"); but how long did it take for managers to stop writing reports in longhand and get them typed? How many still do? Agent technology, digital TV (when and if it arrives), and further integration of telecommunications with transactions (as enabled by voice recognition—not just "press 7 to hear this confusing menu yet again") will create other content models for retail. Imagine watching a video at home, liking a particular pair of jeans on some actor, and clicking on them to bring up the Web site with a distributor. Now imagine sending a buy request to a virtual jeans market.

Airline tickets are a particularly interesting electronic market example, because the way they're sold is approaching the sophistication of financial instruments. Customers can purchase an "option" on a ticket (i.e., make a booking that can be canceled later), or purchase a "future" (i.e., buy the ticket early, on the expectation that the price will increase). Hence market mechanisms are becoming far more sophisticated than the typical retail operation, especially when you realize that tickets are a perishable good: an empty seat is an opportunity loss. American Airlines maintains a E-mail service where subscribers are E-mailed offers of weekend specials every Thursday. Over 100,000 subscribers are now registered. Other airlines have followed their lead.

TRANSACTION PROCESSING

The most basic exchange between customers and retailers on the Web is on-line ordering. A virtual catalog can include an on-line order form where a visitor can complete an order, provide shipping details and a credit card number, and then just push a button. The difficulty, up until now, has been the payment aspect of this. Aware that consumers are still worried about payments via the Net, retailers have devised a number of ways to ease the worry. One is to provide a version of the order form that can be printed out and then faxed. A better solution is for the shop to call the customer back to get a credit-card number over the phone.

Virtual malls obviously have a different transaction challenge—they have to manage transactions for each shop, allocating orders and payment appropriately. One of the reasons to place your site in a mall is to have these services provided. Virtual malls have, in some instances, been able to finesse payment problems by having consumers register with a credit card number. Thus, the consumer trusts the mall to accurately bill the credit card. This is also what the on-line services do, and AOL, Prodigy, and CompuServe can be considered as large virtual malls since they contain numerous third-party retail services. A number of these, such as 1–800-FLOWERS, have done very good business.

The next level of necessary transaction processing is probably the "shopping basket." A virtual store where a visitor may want to purchase multiple items has to provide a virtual analogy to the shopping basket, where a consumer can place items (and remove them) prior to purchase. The technology to do this is not complicated—the Web server has to be able to collect a list of items, and allow people to review the list and remove from it. What is amazing is the number of stores that expect people to buy without provision of this

facility; it's as if someone opened a grocery store and re-
fused to provide baskets or trolleys. On-line services suffer
from the disparity in the ways that their various stores ask
their customers to place orders ("Want to buy one? Please
fill in this form making sure you get the stock number you
saw two pages ago exactly right").

Virtual bazaars are in a rather different position when it
comes to transactions: Their members are predefined. The
bazaar offers a wonderful opportunity for *reconciliation*, basically agreeing to pay the difference between trades. Companies in certain areas have done this since time began, but
electronic bazaars offer the possibility for dynamic accounting. Imagine a bazaar where publishers can buy, sell, and
trade written content. Publisher A, wanting to put together
a specialist report for a client, can bid on material owned by
other bazaar members, who can then get paid or can agree
to some possible future use of material owned by Publisher
A. The possibilities are considerable.

Theoretically, a virtual brokerage could, on receiving a request, get on the phone and generate insurance quotes. But
this is patently ridiculous—*real time* operations require integration of the brokerage with the supplier's systems, as
shown in Figure 7–2. This may be as simple as sending a
database inquiry and receiving a response, but in the future
the interaction will be more complex. If the customer has
complex requirements ("I'd like coverage for X, but only if
less than $200 a year"), then brokerages will have to develop
iterative procedures, somewhat akin to refining a search
when using a search engine. Alternatively, customers will
have to be given the opportunity to refine their requests.

Sometime in the future, a virtual brokerage will deal
with a fluctuating number of suppliers. Requests for quotes
will be dispatched, perhaps to newsgroups and other third-
party areas, and various responses will have to be compared
and put into standard format. In a way, dynamic brokerage

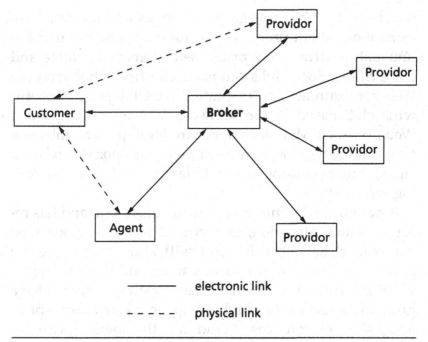

Figure 7–2. Electronic Real-Time Brokerage.

will move toward the agent model. Now that's something that we'd call an electronic market.

CONSUMER POWER

Less developed in Web-based retail are the tools that *consumers* will use, modify, and perhaps even develop for electronic purchase. Unlike physical retail, it cannot be assumed that consumers will just park their cars and walk into the store. They have other options; they can use software to help them.

One model that is well grounded in physical shopping is the notion of *clubs*. People come together in clubs to purchase items where the transaction costs make individual purchase prohibitive (e.g., some stocks and bonds), or to learn more about products and have options to purchase unusual or attractively priced selections (e.g., coffee and wine clubs). Such clubs can naturally find a home on the Web; for example, The Tri-State Wine Club is a (for profit) wine club based in the Midwest with a well-developed Web presence. Manufacturers can develop such clubs as a way of getting closer to consumers, or sponsor existing clubs, thus bypassing Web retailers as sources of marketing information.

A paradigm that has no physical equivalent, and has received much attention, is *electronic agents*. Anyone who has read closely to this point will know that we're big fans of *agent technology* as a way to extend the functionality of the Web. Agents can be developed that search for a product or service to purchase, given a consumer's price and other preferences. "Find me the lowest price for Mozart's *Jupiter* Symphony" or "Find me everyone who's selling Donna Karan evening dresses" are appropriate tasks. Bargain Finder is a prototype agent developed by Andersen Consulting that finds the lowest price on a specific compact disc. Nine Lives Clothing Store, discussed in Chapter 1, showed how a simple agent can be used within a single shop.

Agents pose a challenge for the virtual store. By reducing shopping to a search mechanism, they eliminate the feel for service and reliability that a visitor can obtain by browsing. Bargain Finder is actually blocked by most of the stores that it can visit; the vendors argue that its price-based comparative mechanism is a gross simplification of the shopping process (and, to be fair, they have a good point). Despite this, the extensive use of agents in the future to at least narrow the search, or perhaps search out alternative outlets, seems certain. Use of agents may place considerable power in the hands of the electronic consumer.

In the future, Web retailers may have to arrange their Web pages and offerings to attract agents, not human shoppers. This movement can be seen in some existing stores that arrange their pages to achieve many search engine "hits." One trick is to hide words on the page that people often search for, even if the product has nothing to do with those words. Thus, a seller of coffee might hide the word "sex" on the page (hiding text is easy—simply specify the background and foreground color to be the same).

INTERACTIVE MARKETING

Well organized virtual stores do not just rely on pure content to attract customers: They adapt it for their own use and provide associated facilities that result in true *interactive marketing* between consumer and retailer before, during, and after the sale. This is perhaps the true power of Web-based retail—putting it all together.

The most common tool in interactive marketing is giving consumers direct access to human know-how. Kaleidospace allows visitors to E-mail the artists about the work they are displaying, and have reported that most purchasers of art have an E-mail conversation with the artist first. Virtual Vineyards gives tips for wine selection by a wine expert and allows visitor to E-mail the expert. The responses are collected into a FAQ (Frequently Asked Questions) file that is often a convenient way of providing deeper information that can be driven by consumers.

Like regular physical shopping, free samples and trials are a popular part of interactive marketing. Software.net provides lots of shareware; Oracle lets users download versions of software for a 90-day trial period. Record companies like Delos International (a small producer of specialized classical music) provide audio clips of their discs, and larger companies have set up entire sites for a single disc that contain clips or even extra tracks not available in store. In the future, one can image virtual coupons that can be used at virtual stores, or printed out for physical stores. There is already a physical mall based in San Diego (the UTC Mall) that provides a Web site which can be used to print out physical coupons for the mall.

Mass Customization

In some incarnations of interactive marketing, a consumer or client can interact with the company to shape the service or product. Such personal marketing has existed for as long as there has been commerce, but until recently has been associated with high-priced goods (e.g., boats or automobiles) or goods that can result in fee income for the seller or an intermediary (e.g., stock purchases or life insurance). The customer can inquire, specify, and "sign off" on the service or product, and do all of this within a limited time frame determined by the customer. Ideally, the customer can also make payment electronically. For information or digital products, the service or product can be distributed to the customer over the Net. Where distribution is not possible, enough imitation material and information can be provided to have the consumer accept the physical product.

Software, financial analysis, textbooks, and electronic art are examples of areas where this level of interactive marketing can now be achieved. For example, ESI (based in England), E*net, and other discount brokerages allow subscribers to receive real-time stock information and make stock trades. Stocks can be traded with the physical identification of the stock changing hands after the time of the deal. Publishers can now customize textbooks for specific courses, and there is no reason why all or part of these cannot be delivered over the Net. Consumers can browse and agree on an imitation (in this case an electronic version) of the text and may not need to see the physical copy prior to purchase.

Small start-up companies have provided fascinating examples of new variations of products based on this highly interactive approach. Build-a-Card lets a user build a greeting card at its Web site; the card can then be made available as a Web page to the receiver. While rather simple at present, technological advances may make this and other efforts the preferred way of sending cards. (The likes of Hallmark have not been slow to notice the potential for electronic greeting cards.) The Cyrano Server writes love letters; the user specifies the basic elements of a letter, which is then crafted by the service and E-mailed to the user. (The service did E-mail letters anonymously for users, but has had to discontinue this.) In an interesting example of learning about markets, the Cyrano Server has found that there is a considerable demand for letters that break off relationships.

The sale of personalized apparel can be managed through this interactive approach. A consumer can create a graphic design and submit various versions of it, see pictures of simulated shirts with the designs on them, choose a design, agree on quantity and price, and then wait for the printed shirts to arrive. This can be done now with some apparel specialists, but the Web allows this process to happen in a very short time frame between parties who may not even be aware of each other's physical location. Examples of small apparel businesses providing this service include Graphiti and Namark.

ON-LINE RETAIL GROWTH

If the Web is a marketplace that retailers can sell into, the critical question is "How big will it grow?" Is Web-based retailing the next big move, or just a minor outreach? The answer to this, as might be expected, is not simple. Our opinion is that the Web will become an important delivery mechanism for some products and services, but not for others.

A number of consultancy groups have monitored or surveyed on-line retail. The Forrester Group has estimated that on-line sales volume in the United States in 1996 will be about $518 million, including $140 million in computer products. However (and this is interesting), only $46 million or so is sales by major retailers in areas such as apparel. By the year 2000, they estimate that major retail sites will grow to be about $6.9 billion, far less than the $46.2 billion in deposits they estimate will be self-managed by consumers on-line. To put this in perspective, this would make on-line retailing less than three times the size of home cable shopping now, and only about 15 percent of the sales of Toys "R" Us. And Wal-Mart is over four times the size of Toys "R" Us.

Other studies are more optimistic. Jupiter Communications has estimated that in the year 2000, U.S. on-line retail will be a $73 billion business, more than 10 times the Forrester number. Some hype-merchants have forecast in the region of $200 billion. But even this figure is less than 2 percent of all U.S. retailing. *The Independent* (1996) in the United Kingdom reported a Hoskyns study that estimated the United Kingdom on-line retail market at £21 billion (about $33 billion), which seems wildly optimistic.

Whichever figures you believe, two important lessons arise from this analysis. First, much of the growth at the moment is coming from start-up companies. A million or two may be a rounding error to Wal-Mart, but for a small company offering a niche product to a global market it may mean a nice little earner. The opportunities are for the small and the nimble. If there are customers on the Web for your product, and you can serve them, go ahead. Second, the products and services that will sell on the Web may be clustered in a few small areas—travel, entertainment, ticketing. Where software can be used to search for something ("find me a vacation") or deliver it ("send me an electronic ticket"), then, as discussed throughout this book, the Web will become an important channel to reaching the mass retail audience.

Pick a Number

We think that the amount of on-line retail in the future is likely to be closer to the Forrester estimate than those of others. However, simple numbers like this ignore three possible scenarios for many firms. First, particularly for the small firm, a Web presence may lead to sales through other channels. Customers may not buy through the Web, but they may be encouraged to order a physical catalog, or call and talk, or pick up a copy of something at their local store. The final sale may be through another channel. We came across a small company where a Web site was one of 28 methods of moving its product (distributors, company store, own physical catalog, entry in others' catalogs . . . we forget the rest).

Second, the Web presents an opportunity to obtain new customers. Many existing sales relationships are fairly static, particularly in the business-to-business area. On-line retail may have an effect on the distribution of customers, with those who prefer or choose on-line shopping gravitating to on-line companies. Third, as discussed in Chapter 2, the simple existence of on-line retail will open up possibilities for new products and services as yet unthought of. For example, it is apparent that interactive games run over the Web may be the next big growth area in the computer game business. There are opportunities to retail the games over the Web by providing trial periods. Whether these new business opportunities turn out to be a minimal part of the picture or result in one or two new retail giants that grow on the back of novel ideas or delivery is yet to be seen.

BEST PRACTICE

The following businesses are interesting examples of electronic retail operations.

Software.Net

Based in California and related to McFee Associates, the virus checker people, Software.Net sells software over the Net. Out of 20,000 or so products that they carry, over 1,300 can be downloaded electronically. Using proprietary algorithms, the software will not install and function until a key provided by Software.Net is used to explode the compressed software. With well over 2,000 orders per week from all over the world, including about 8 percent of their sales from Japan alone, the company continues to grow. They have moved into corporate accounts, and signed a deal to become the first people to distribute Microsoft products electronically.

Apart from this, and a very well-designed virtual catalog, what makes Software.Net really interesting is two other parts of their strategy. First, electronic distribution is really just a means to a strategic end—low-cost software. By removing physical packaging and distribution costs, they can provide software at highly competitive prices. Second, Software.Net continues to distribute lots of shareware. Anybody can ask them to catalog and carry shareware, and if it is suitable they will do so. Thus they become a repository for both free and priced software, generating customers who crossover from shareware.

As might be expected, the Software.Net model is now being copied. Stream is a company that also provides for downloading of purchased software. Its value-added strategy is to provide deep technical support to customers, including discussion groups as well as direct electronic support.

OnSale

Based in California, OnSale was an early innovator in the use of electronic auctions. The company, self-described as a "live on-line auction house," sells used computer and other electronic equipment through auction mechanisms, including traditional highest bid and Dutch auctions. (Its definition of a Dutch auction, however, is very different from the expected definition; unlike Dutch auctions, the price is not lowered over time.) The "Yankee auction" format is also used. This is simply a highest bid auction for multiple lots where the highest X bids gets the available X lots. The following is an example of what the potential bidders see in a Yankee auction: more than enough information to generate a bid that will be in the top five (the quantity available). Also, note the bid from County Cork, Ireland; this is a global auction:

Minimum Bid: $199.00
Bid Increment: $ 15.00
Quantity Available: 5

Auction closes at or after Fri Sep 27, 1996 9:17 am PDT.
Sales Format: Yankee Auction(TM)

Last Bid occurred at Thu Sep 26, 1996 8:38 pm PDT.

The current high bidders are:

1. GH of Sheboygan, WI, Wed Sep 25, 5:03 pm ($319.00, 1)
2. DG of Vancouver, WA, Thu Sep 26, 8:38 pm ($304.00, 1)
3. LS of Ballincollig, COUNTY CORK, Wed Sep 25, 10:46 am ($289.00, 1)
4. JB of St Joseph, MO, Wed Sep 25, 12:51 pm ($289.00, 1)
5. AC of Pittsburgh, PA, Thu Sep 26, 10:16 am ($289.00, 1)

OnSale also provides facilities for customers to monitor and analyze present and previous bids; this is called an "Account Activity System."

InsWeb

InsWeb is an electronic insurance brokerage. They provide on-line quotes for medical insurance from large insurers such as ITT Hartford and Blue Cross/Blue Shield of New Jersey, and on-line quotes in specialized areas such as wedding insurance. (Insuring against the risks associated with the event, we presume, not realizing you've married a loser.) Customers can also be connected to insurance agents to whom quotes will be forwarded, and customers can use a search facility to find an agent.

InsWeb is interesting because it is a substantial attempt to build the insurance industry equivalent of an airline reservation system. Customers will be able to generate and compare quotes in real time. To date, it has been slow to develop new insurance areas (there are lots of nonactive links on its pages), and its eventual success or failure will be of considerable interest.

Wal-Mart

Who would bet against the world's biggest retailer not also being a giant in the electronic world? One of the first buyers of Netscape's commercial server, they now have some of their inventory on-line and delivery available in the Arkansas area (their home state). There are some hints that they are already very Net-savvy. Their Web site, shown in Figure 7–3, portrays the same low-cost image aimed at lower income groups that permeates their physical retailing.

What's interesting about Wal-Mart on the Web is the extent to which they will bring their world-class ability to employ logistics to bear. The way they integrate delivery to the customer, store and electronic sales, and perhaps things like customer pickup of preordered items will be examples for all.

Sainsbury's

Sainsbury's is one of the largest food retailers in the United Kingdom. They also own the Texas Homecare (not based in

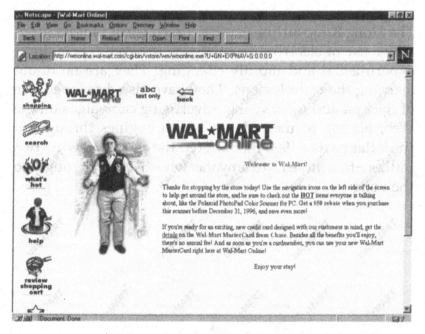

Figure 7–3. A Different Image.

Texas) and Homebase chain of hardware stores. They have interests in U.S. food chains, including outright ownership of Shaw's supermarkets. On the Web since 1994, they have experimented with the Web as both an advertising medium and a direct sales channel. To our knowledge, they were the first people to put an *entire* TV advertisement on the Web. (We won't mention download times.)

Starting with direct ordering of wine, they now provide direct ordering of flowers and chocolate as well. Some "overseas" delivery is possible. Sainsbury's is of interest because the company is slowly extending into direct sales and multinational retail. Distinctly British, their Web site will reflect the extent of their global aspirations, and if they determine that the Web and direct sales are vital components, then no one should bet against them being leaders in this area.

Cathay Pacific

The Hong Kong airline was discussed earlier for its innovative use of electronic auctions on the Web. Their site is experimental and rapidly changing: They are no doubt learning the right lessons. They have also developed one of the best and most visible advertising campaigns on the Web, placing banner ads on search engines throughout the latter part of 1995 that offered the chance to win "two million free miles" to anyone who clicked through to their site.

8

Business to Business

Consumer products are sold to the masses, the unidentifiable crowd yearning for better mouthwash. Goods are sold through a combination of mass marketing, brand loyalty, price, and selection. The business-to-business crowd, on the other hand, usually knows their customers very well and seeks to establish long-term relationships, if for no other reason than to wait out the "Net 30 days" terms of the invoice (consumers, on the other hand, pay at the register).

By now your head is full of the promise of one-to-one marketing. You, the seller of goods, knows everything there is to know about your buyer. You watch her on your Web site. You track her every move. You tailor your offering to her every whim. And rather than being arrested for stalking, you are rewarded with a customer for life.

That may very well be the promise and no doubt, by the time this book hits the market, will be implemented in hundreds of Web sites the world over. Dozens of companies will ride the software that makes this possible to successful IPOs and gads of newfound wealth. Or we may have to rethink the nature of the Web and information based marketing, and revisit the basics.

THE 5 Ps

In the beginning there were the 5 Ps: People, Product, Price, Promotion, Distribution. (Sure, "Distribution" starts with a "D," but it's considered an honorary P.) These marketing basics are what we learn in Marketing 101. They provide the foundation from which all other marketing springs (see Table 8–1). So why then has many an otherwise smart marketing company thrown out the basics on the Net? To answer that, we need to review the basics, look at some examples, and return to our roots.

Table 8–1. The 5 Ps of Marketing.

The Ps	Description
People	Users of your Web site, in particular your business-to-business users
Product	Defining new uses of information and changing business models
Pricing	Micropayments, on-line transactions, and the like
Promotion	On-line ads, games, and introductions
Distribution	New ways to increase your distribution base
Presence	Designing for the business-to-business user

The Web is a different animal and requires a different viewpoint in business-to-business marketing. We have developed the sixth P, *Presence*, to help us understand some of the roles that a Web site contributes to the marketing mix.

AN OVERVIEW

People

Knowing your customers better is one of the big attractions to doing business on the Net. As already pointed out, you can do a great deal of work in analyzing your log files to see where the hits are coming from, you can issue cookies to track a user through your site or track repeat users, and so on. All of these things are useful, but what can you do to better know your customer?

The problem with a Web site is that you never know who will visit. One minute, it is an important customer yearning for knowledge, and the next it is some wise guy college kid sending you rude E-mail. Such is the nature of the neighborhood.

There seems to be a split on the Web as to which of two approaches to take; a simple presentation of the company with little technical information but lots of good-looking graphics, or serious information targeted at your customer base.

Let's look at the serious sites first. By now you have been GE Plastic'd to death, but they began as and remain one of the stellar sites in providing detailed technical information to an audience searching for detailed technical information. Enough said. But the site speaks volumes about what GE knows about their customers. They have put the specifiers, designers, molders, and extruders of plastics first on the list and give them the most meat. The commercial divisions in the United Technologies Corporation Carrier site do the same thing. Detail means that your customers know that you have made a commitment to the medium and that they can expect to receive good information from using your site. Remember, in the end you do a site for your customers; give them what they need.

UTC Carrier's Comfort at Work

Carrier's commercial division makes very large heating, ventilation, and air conditioning units (HVAC). The systems that they build are the kind that heat and cool entire buildings or even multiple buildings. Their audience for information is a varied lot, ranging from architects who may design a building with a particular system in mind, to administrators assigned the task of finding replacement parts for an existing system. With such a wide range of potential users, you can assume that the expertise level varies as greatly. Some have an intimate knowledge of the HVAC industry and years of hands-on experience; others may not know where the off switch is. In a non-Internet world, serving these varied customers was fairly straightforward. Teams of salespeople would educate, inform, provide information, and close the deal. But the world of the Net changed all that because now these people can come for information unannounced and expect to receive that information unattended.

The Carrier team handled the problem like a matrix. In one corner, you have people who are unfamiliar with HVAC and unfamiliar with Carrier products. In the opposite corner, you have people who are very familiar with both HVAC and Carrier products. The information required by these two groups ranges from general to specific. As important, the presentation of that information is also very different.

To work these very different crowds, Carrier came up with a three-path approach, named appropriately, Path One, Path Two, and Path Three. Path One was designed for novice users and assumes little or no specific knowledge of the HVAC industry. All you have to know is that you want an HVAC unit. The Path is build around a series of questions presented in a very simple graphical format. The problem of the user interface is solved.

The "eye candy" sites are usually first attempts to get online. These companies think that if they put up some great-looking pages they somehow send a message that they are modern businesses and their customers will derive value from that. Allied Signal is one of those sites. Their site provides some good information for the casual browser, tells you the types of businesses that Allied Signal is in, walks you through some of their product lines. But there is no detail. If I seriously wanted to buy something from them or design something using their products, I cannot do it from the Web. Missed opportunity.

Sometimes your end user is not your target audience at all. Sometimes the audience is your investors. Take a look at the Novartis site for a great-looking, detail-shallow site but one that you leave thinking, "Hmmm, if they can make a site this cool, maybe I should invest that inheritance." (Novartis, if you recall, is the merged company of Ciba-Geigy and Sandoz.) Even though you can barely find out what this company does in any detail, their customers are definitely not the target. Different audiences require different styles.

Presenting the Single Face

Multinational conglomerates are great things. Imagine that from a single company you can buy helicopter gunships, elevators, and air conditioners, or from another dishwashers, credit cards, computers, and jet engines. These conglomerates face many and varied issues in conducting their business, not the least of which is how to create a way for customers to buy from each and every facet of their businesses. Creating this "single face" to the customer is usually one of the very first reasons for building a Web site.

In developing marketing literature, teams of marketeers usually come up with a simple way to present a company: by products and services, by division, sometimes by function. These classifications usually take on a life of their own until groups like Company X Information Technology becomes CXIT or Brand D Appliance Division becomes Bapp or some other corporate speak. While this may work inside the company, customers usually have a harder time catching on and only become confused when told that you cannot sell them their light bulbs because that kind of sale is handled by MYOP. Silly stuff for sure but we have seen this kind of thinking in every organization that we have visited. This makes the job of a marketeer increasingly hard and very difficult to convey on paper and in print. What usually happens is that after the groups are defined, they embark on separate paths in creating their look to the world. Brand D chooses a purple motif with a Times Roman font for it's logo, while CXIT goes black and white. When you put all these brochures on a single table, the differences become all too apparent and you begin to wonder if it is indeed one company after all.

The Web offers a solution of sorts and it is one that many companies intuitively realize. Take a look at Westinghouse, one of those industrial giants that has been on the Web since 1994 and has been expanding into new industries. Westinghouse decided that its groups lined up neatly into an Industries and Technology Group, its CBS Group, and some basic corporate information. While power generation is a long way from CBS broadcasting, while you are on the Westinghouse site you would never know (see Figure 8–1).

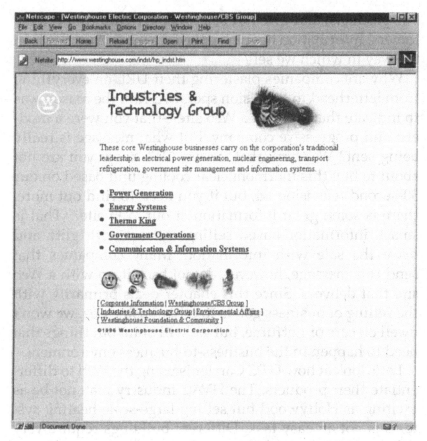

Figure 8–1. Consistent Image Is Everything.

Each area is color coded and the layout is the same. Navigation is simple and straightforward and each of the groups has a fair amount of material. If you wanted to buy a satellite and a steam turbine, you would know that Westinghouse made both.

Product

What sells products? If you guessed information, then you can skip to the next section. If not, read on. Information is what we use to create our marketing materials, differentiate our products, and give our salespeople to inform our customers. Information has never played a more crucial role in the way in which we sell.

Why are companies plastering their URL on everything from letterhead to television spots? At first the reason was to indicate that you had a Web site—that you were a modern and progressive company. But what message is really being sent? We think that it says, "We know you are not about to buy this thermonuclear cooling unit based on our 30-second television ad, but if you need to find out more, there is some great information at our Web site." That is smart, information-based selling. Attract with glitz and drive the sale with information. Many companies that send this message, however, do not back it up with a Web site that delivers. Since this chapter deals primarily with the selling of business goods to other businesses, we won't dwell on cars or perfume, but we will point out things that need to happen in the business-to-business environment.

Let's look at how UTC Carrier is using the Web to differentiate their products. The HVAC industry may not be as exciting as Hollywood but selling large-scale heating systems is not an easy task. Different buildings require very different HVAC systems most of which are quite complex, requiring the matching of hundreds of components.

So how do you sell a system like this? Certainly you do not walk into your local HVAC store and expect to walk home with one, but local dealers and experts are the folks that close the deal. Carrier thought that there was a significant opportunity to start the sales cycle with people who may be responsible for deciding brands and system types without necessarily being knowledgeable in the HVAC arena. To reach these specifiers, Carrier produced an elaborate Web site that walks a novice user down what they call Path One. The choices are presented with a visual navigation system that lets you know where you are in the system.

Carrier sells many of its systems through specifiers, or people who are responsible for planning the system and defining the system requirements. These folks usually are not the same people who cut the check, but they have the final say in which system gets chosen. But how does the specifier use the Web? By choosing a generic building type, like School or College, the user is presented with systems that can be used to heat and cool that type of structure. A quick click on one of the building types brings up a description of the system and the chance to gather more information about the actual system components. There are numerous components for each system but many of the components overlap. If we choose to go on, we can gain detailed product specifications and tables that give us the highest level of product information. This information is contained in an Oracle database that receives updates as the information changes. Everything is up to the minute to provide the user with the latest information 24 hours a day, 7 days a week.

We usually think of products as tangible objects, things we can hold, see, feel, for want of a better term: widgets. Simple right? But what about services? Traditionally, services support products; we build a sales support staff to back up our salespeople and technical support, and help desk people to answer questions about the products, marketing folks to create demand, and accountants to tell us how much the products cost and how many widgets people are buying. These services are all too often given away because of pricing pressure within an industry. For example, it is still common in the computer software industry to give away phone support for certain products. In the value chain, these are known as support services. They add to the cost of the good, but the price of the good rarely reflects their total cost. The more complex the industry, the most support is required and the higher the support overhead.

Many companies are turning to the Web for their base level of customer support.

Consider Cisco, maker of the many of the routers that move information across the Internet. A router is a specialized, high-speed computer with a simple task, move the right packet of information to the correct address in the correct order and make sure that none of that information is misplaced. If only all business were so simple.

The challenge that Cisco faces is that as the Internet and intranet market continues to explode at exponential growth rates, more people are buying their routers. The problem is compounded because Cisco is no longer just selling to Information Systems groups with years of wide area networking and Internetworking experience. Now any two guys with a T-1 line and space in a dorm room can become an Internet Service Provider. So the experience level of many of Cisco's customers has gone down while their demand for support has increased. A tour of the Cisco Web site is a tour of a site designed to provide very high levels of customer support (see Figure 8–2). In addition to the standard Internet software available for download (i.e., patches, bug fixes, and the like), Cisco helps users self-diagnose their problems. Customers can, through a series of question-and-answer hypertext links, easily provide themselves with common answers to common problems or be directed to the correct area of the company for further assistance.

The impact of this site is far-reaching indeed: If Cisco can reduce the number of calls to its technical support staff by 10 calls per day, and assuming that the call takes an average of 10 minutes to answer (very conservative assumptions for this industry), Cisco will have saved approximately 15 weeks of a worker's time per year. Add to this the savings of 15 weeks of 800-number charges and reduced stress levels, and you can start to see where the big savings can occur.

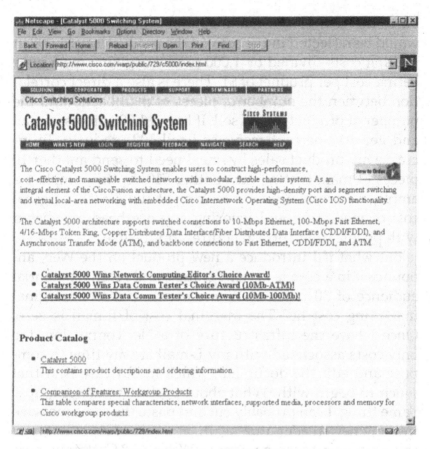

Figure 8–2. Cisco's On-Line Product Catalog.

Price

How Pricing Is Affected by the Growth of the Web

The simplistic way to arrive at the price of a product is to add the costs that go into the product, such as materials, labor, a contribution to overhead, and then a percentage needed to satisfy profit and there we have it, Price.

Theoretically, the costs associated with marketing a product are included in the overall price of the product. For example, if I introduce a new product with a direct-mail blitz, the cost of the mailings and associated follow-up would be reflected in the price of the product, say $1,000 in mailing costs divided by 1,000 products giving me a marketing cost per product of $1. There is also a direct correlation between the number of pieces of mail I send and the number of products that I sell. If I send 1,000 pieces of mail and get a 10 percent response I sell 100 products. To increase my product sales by one, I need to send another 10 pieces of mail, costing me another $10. In this simple example, we are really only looking at the direct marginal cost of marketing and ignoring any overhead associated with people time, infrastructure, and the like.

But what if I introduce a new product on the Web, announce it in a couple of newsgroups, and reach a potential audience of 30 million? Where does my direct marginal marketing cost lie? The marginal cost of E-mail is zero. Once I have the infrastructure of a Net connection, the only costs associated with my E-mail are my time to compose and edit the document, which in our case isn't that much to begin with. What about posting to newsgroups? Same thing. I can probably cut and paste the same message into my newsreader and, again, at a cost of zero. But, you say, what about turning it into a Web page? Certainly more work is involved in this case, but again it is all manual labor and part of overhead. So to push my message into three different electronic media, my marginal cost is zero.

How does that reflect on my pricing strategy? In most business-to-business cases, prices vary by the channel. GE Plastics, which deals in orders in the thousands of pounds of plastic, has one price for customers that buy direct. Since not everyone needs 10,000 pounds of Lexan sitting in their garages, GE also sells through a distributor and gives another price to them. Ultimately, the smaller end user pays a different price than the large end user. Most goods that go through a distributor or warehouse work in a similar manner. We should focus on the direct selling method, since many small to midsize businesses view the Web as a way to increase their direct visibility.

Our $10 item costs us $1 to market via direct mail and zero dollars to promote on the Net. Therefore, a customer who purchases as a result of receiving direct mail should pay $1 more than the customer who purchases from a contact on the Net.

Smaller Units, Smaller Price

The software industry is in the midst of a radical transformation because of the changes in pricing brought about by the very model just described. Marginal cost in distribution of software has always been very low since the price of replicating a floppy disk or CD ROM is in the pennies per thousand. What held the prices high has been the cost of distribution, technical support, marketing, and documentation. The costs and methods of distribution are discussed later in this chapter. Earlier in this chapter we touched on Cisco, so no need for repetition here; documentation is being handled in the same way. If you want a copy of the manuals for Netscape Navigator, you call up Netscape, place an order for them, and pay. If you want the Web version, go over to their Web site and have a look. Netscape's marginal cost in having you view their documents on their server is near zero. Notice how often that zero figure crops up in talking about the Web? Marketing we have already touched on (in terms of pricing at least).

But bringing the costs associated with distributing, marketing, supporting, or documenting to near zero is hardly revolutionary. These changes merely represent improvements in the underlying processes. The true revolution is happening in the software being sold.

The desktop is dead for people in the software business. Think about the applications that you use. They are most likely a word processor, spreadsheet and presentation package, maybe a page layout program, maybe a personal organizer but in terms of non-networked personal productivity applications these are probably pretty much it. And regardless of your beliefs on what the best of these applications are, Microsoft pretty much owns the market.

The network is alive and kicking. The browser is one of the few new applications to become a hit that everyone needs to have, E-mail is on everyone's desktop, newsreaders are included in the browsers, the best server software is free (shameless plug for Apache!), and the development community is thriving. And aside from the big players fighting the browser/HTML editor wars, there is more network-oriented software being developed today than ever before. This new breed of software is called *object oriented*, meaning that it can be used by other people and other programs to build more sophisticated applications. The people who are buying this software tend to be people inside corporations, for use in developing internal applications. And where are they buying these things? On the Net.

Promotion

The Giveaway Strategy
One of the dominant strategies today in the software industry is the free distribution of products. Look at all the major players and they are all giving something away. What immediately springs to mind is the saying, "We may be losing money on the sale, but we'll make it up in volume." How do these companies avoid becoming part of the punchline?

The giveaway strategy calls for the establishment of standards, which if adopted, should give the company that sets them a dominant position since they control the way in which the standards are implemented. The classic example is Microsoft, which has established its Windows operating system as the standard operating system for PCs and is trying to establish WindowsNT as the standard for networked computing. The idea goes something like this: If everybody uses my software when I give it away, then at some point in time everyone will have developed such a dependency on my software that when I start charging for upgrades that they will have to pay and I will be able to get rich. Sacrifice today for profits tomorrow.

McAfee is the pioneer company of the "give away today, make them pay tomorrow" strategy. Long before Netscape was a company or Microsoft gave away a browser, McAfee used the Net to freely distribute its antivirus software. In fact, it was distributing software this way before the birth of the Web. A computer virus is a destructive program, intentionally designed to do harm to your computer or computer network. A few years ago, a virus called *Friday the 13th* made corporations around the world sweat in fear of losing reams of corporate data. McAfee engineers knew early on that network computing made the spread of viruses easier and more prolific. With a network connection, determined virus creators could infect thousands of computers without having to infect and spread a single diskette themselves. McAfee was also a small company at the time and needed to find a way to get software into the hands of as many people as possible for as little money as possible. To add to its problems, the makers of virus software created new strains and new programs that meant the McAfee software needed to be constantly updated to guard against the ever-changing threat.

McAfee turned to the Net to solve both price and distribution problems. Potential users were allowed to download evaluation versions of the software for a 30-day free evaluation. Once the 30 days had ended, a timer within the program displayed a message that urged users to pay their licensing fee. The software never stopped working, just provided this gentle reminder. Once the users had paid for their copy, they received access to another server that would give them complete and registered copies of the antivirus software. A paid license entitled you to download updates as often as needed. McAfee would warn users of new viruses on the Net and via E-mail messages, ensuring a continued stream of new users, product upgrades, and new evaluation users. Today, from these humble beginnings, McAfee is a publicly traded company with revenues totaling $181 million in 1996. Building on this success, McAfee also formed Software.Net, identified in Chapter 7 as a leader in Net-based software retail.

Designing for Promotion
There are two reasons to design a Web site. The first is that you are trying to communicate your message to more people than you could through other channels. The second is that you are trying to reach fewer people but give those people more information than you could through other channels. The first reason is why most people put up a Web site.

Too often, we see Web sites falling into the trap of trying to be everything to everybody. A good promotional site *targets* a particular theme and a particular audience. The *scope* of the site does not try to overreach, but rather pays close attention to the information needs of the group for which the site is designed. How then, do we design a site for promotion? Simple. Target an audience and forget about the other 30 million people on the Net.

A good example is the Tide clothesline site (Figure 8–3). This is one of the Procter & Gamble sites but other than in a disclaimer at the bottom of the home page, you would never know it. Is there a link to Investor Relations, a message from the president or the Annual Report? How about press releases, hiring information, or directions to the company headquarters? No and no again. The reason you will not find any of that is because Tide is a site designed to promote Tide and nothing but Tide.

What you will find is an area devoted to Tips and Timesavers providing washing advice (Figure 8–3). The Stain Detective is a perfect example of the interactive nature of the Web. You input your type of stain, answer a few questions, and you have a customized analysis of how best to remove that pesky gravy stain from your favorite tie. There are also areas for playing a game and winning prizes, some tidbits about the Tide-sponsored NASCAR race team and a store to buy Tide-related merchandise. Very targeted, very effective.

Banner Ads

Perhaps the easiest concept for an advertising exec. to understand is the banner ad. These small rectangular ads appear on the tops of all the popular search engines, many of the on-line magazines and even on some of the software vendor sites. Banner ads look like print ads. But a banner ad takes advantage of the hypertextual nature of the Web by allowing the user to click on them to gain more information. Clever advertisers may even link to a special place on their Web site that only people who click on the ad can reach.

Much of the money spent on advertising on the Web goes into these banner ads. An entire banner ad industry is growing, led by companies like Double Click and the Internet LinkExchange (ILE). Double Click was started by some

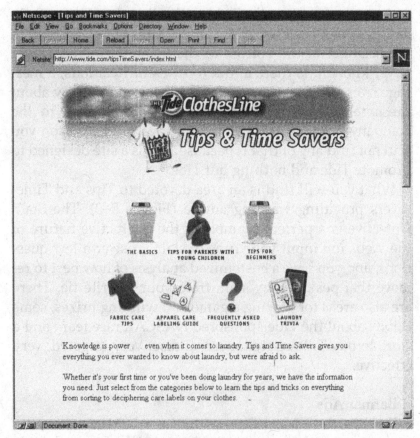

Figure 8–3. Tide Cleans Up On-Line.

ex-advertising people to provide a type of banner ad clear-inghouse to sites that otherwise might not have the pull to appeal to national advertisers. Double Click promises to target the ads to the demographic chosen by the advertiser, not by the publisher. From a central bank of servers, Double Click passes their banners across the Net to the awaiting public.

The Internet LinkExchange is a much more grassroots effort. Prompted by what they felt was an increasingly commercial use of the Net, the people at the ILE decided to

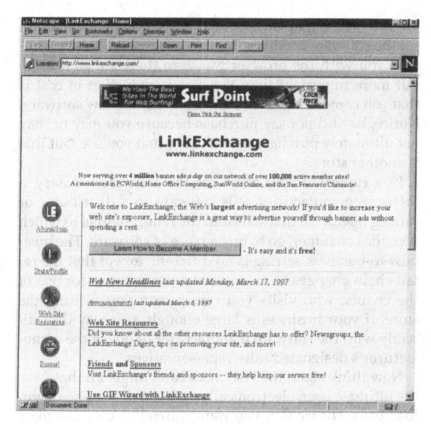

Figure 8–4. Scratching Mutual Backs—The Internet LinkExchange.

simply ask the people who ran sites to provide free links to their network of advertisers in exchange for a free ad on the network. This appeals to smaller advertisers since they do not have to put up large sums of money to join the ILE network (see Figure 8–4).

Distribution

Where did your browser come from? Unless you are in a company that pays for a large volume site license, chances are you picked it up on the Net. This is a very good method for the software company because its marginal cost to provide you with the browser was zero (here we go again!). But more important than the drastic reduction in cost is that you came to the site as a place to gather new software. Notice, we did not say purchase because you may or may not ultimately purchase the software that you use, but that is another story.

In a typical supply chain for the software industry, a software manufacturer sells software to a distributor or catalog operation that sells into the retail channel to which you, the consumer, go to buy a piece of software. The business-to-business setting is no different, except that the retail chain changes to an account representative for one of the chains, who visits you rather than you visiting the store. If your business is large enough, you may deal directly with the software manufacturer or one of the manufacturer's designated sales representatives.

Now think about the Web model and forget all that. You, the ultimate user, electronically visit a Web site and download the software that you want yourself. No middleman, no retailer, no warehousing of goods, no cost of inventory—just pure bitstreams of profit.

The software example is easy since the product is already electronic, and to cut off here would be a cheap and easy way out of a fairly complex problem. Indeed, if you are a software distributor, you are most likely very worried about how you can keep your business from eroding in the coming years. In an attempt to overcome this problem, some of these distributors are changing their channels to a more direct method of reaching potential customers. The traditional warehouse distributor focuses on keeping the retail channel supplied with goods. But in the case of general consumer distribution, stores like BJ's Wholesale Club and Sam's Club offer to cut out the middleman and give consumers direct access to warehouse quantities and pricing. This undercuts the general retail channel and leads to lower pricing, fewer retail outlets, and so on. Who then is becoming the Sam's Club of the Web?

The business of BusinessWire is the warehousing and distributing of business-to-business news. If you send out a company press release, chances are you send it to BusinessWire. Its traditional role is to distribute your company news to as many trade journals, business periodicals, and newspapers as you are willing to pay for. Editors and writers turn to BusinessWire because they know that they can count on receiving a certain number of stories each day to fill out their papers and magazines. It is also, frankly, easier to edit a press release into a "news story" than it is to pay a reporter to go out and write one. The value of BusinessWire lies in the ability to get your story printed in as many publications as possible.

But as discussed earlier, the cost of Web promotion has driven the marginal cost of promoting products to near zero. You can just as easily add an editor's E-mail address to your press release distribution list as BusinessWire can. And since you are posting your press releases on your Web site, you can reach the mass audience that once required the resources of BusinessWire. What then is the role of the information distributor in the new media age?

The people at BusinessWire have taken a very hard look at the company's role and rather than sit back, waiting for customers to leave, they have repositioned themselves for the digital age. First, they put up their own Web site. For the PR industry, this is a big deal. Newspapers and periodicals were once the only targets that BusinessWire would distribute to. Now the messages of their customers go directly to consumers through their Web site. One of the common practices in any PR firm is to gather competitive information. This used to mean reading dozens of periodicals and trade rags, clipping the stories, copying them, binding them in a neat little notebook, and passing them off to some supervisor who would present them to the client once a month. Now PR hacks (press agents) scan sites like BusinessWire because the releases are posted on the site at the very time associated with the release. No more waiting a week for it to hit the newsstands. Attraction to the PR industry and decreasing time of information to market are two ways that BusinessWire has increased its value.

BusinessWire has also expanded on its traditional value, access to publications. In the print world, if you want something to be picked up by the Associated Press, you had either better have one hell of a big story or, more likely, access to a service like BusinessWire. Having your story picked up by AP means that you will receive national or perhaps international coverage since many papers rely on the wire service to act as a filter and filler. *Filter* meaning that AP screens and edits the stories, and *filler* meaning that if there is a gap on page three of tomorrow's newspaper the editor can fill it with a story from the AP.

In the on-line world, business news comes from more than traditional sources such as newspapers and magazines. You may get your news direct from the manufacturer, from a mailing list, from newsgroups or from one of the many Net-based services. If you are like us, you get some of it from Yahoo! And if you are like Yahoo! you get your business news from BusinessWire. Your company can get a release on Yahoo! in no other way. BusinessWire has access, your company wants access, distributorship preserved.

The downside of this type of distribution is that a press release now holds the same status as a news story. And as anyone who has ever drafted a press release knows, PR ain't always news. By the time a writer reads the release, makes a couple of phone calls, writes a story, submits it to the editor who reviews it and okays the story for publication, it has become old news. This is a serious problem in industries like the software industry, which react swiftly to changes in events. Rather than deductive reporting, too many magazines are relying on press releases to fill the void. The real news behind the release shows up somewhere else, usually in a Usenet newsgroup.

Creating Presence—The Final P

In architecture, presence is very easy to understand. Walk into any cathedral anywhere in the world and you will be immediately struck with an overwhelming sense of, among other things, presence. The same holds true for most large banks, large corporate headquarters, and capitals of government.

In the shifting, changing and evolving world on-line, presence can be an elusive quality. Certainly eye-catching graphics and the style of your Web page can project certain corporate qualities but what if your customer turns the graphics off? Your prose may help, but that tends to be more stylistic than presence related. We define presence as leadership and preservation: *leadership*, in the way your company uses the Web, in newsgroups and on mailing lists; and *preservation* in regard to brand identity.

Leadership—UTC Carrier

If you think of the Web as a single document interconnected with hyperlinks (those annoying blue words, remember?) then what is to stop a company from tying its information to yours? What indeed. In the early days of the Web, it was common practice to point to sites from your home page that you thought might be of interest to other people. These sites might or might not have pertained to your business, they might have been general business tools, pointers to new software, and the obligatory link to the Dilbert site. We did it both to remember where in the heck those things were for ourselves and also to help others who might be new to Web navigation. Indeed, it became somewhat chic to have pointers to sites that no one else had found or to something so bizarre as to be scary. But those days are gone for all but a few, replaced by large search engines like Alta Vista, categorizers like Yahoo! or reviewers like us.

The opportunity for companies within particular vertical markets (as opposed to general business markets or consumer markets) is to create a series of links to all the sites that fall within your industry, thereby creating an information center for the industry. The key is to gather the most comprehensive and up-to-date list. Doing so builds confidence in your suppliers and customers that whatever their problem or need, they can turn to your center for help and resources.

UTC Carrier took this approach when they launched their Web site in August 1996. Carrier competes in the HVAC industry and their major competitors, Trane, York, and Lennox already had developed Web sites. In fact, Trane had already been through a generation of development and a couple of site redesigns. Carrier needed to launch a site that proved to their customers that they were committed to maintaining technological leadership; in short, they needed to leapfrog their competitors' designs and provide more value in the site itself.

Establishing leadership was extremely important to them. The company had, after all, invented the air conditioner. So they set about creating the comprehensive guide to the HVAC world on the Net, their Carrier Resource Center. The Center contains links to industry organizations, universities that conduct HVAC research, trade shows that maintain an on-line presence, and other resources. They also put in things that were unique to Carrier and their customers, such as white papers, press releases, and industry speeches. The important point is that anyone looking for virtually anything in the HVAC business can turn to Carrier for answers and links.

So what did Carrier leave out in its hyperlinks? Pointers to competitors. We think it is important to link to your competitors if you hope to establish true leadership on-line. Let your customers see what is out there. Let them see that your site reflects your company's commitment to its customers. Do not run scared, do not think that the old rules apply. Users will be grateful for the help, and that is what will get them to come back. They will know that your site will show them the way.

Who's in Charge?

Even if your company has been on the Web for more than a couple of years, chances are you are still treading water in virgin seas (to mix some metaphors). One of the vexing dilemmas that each company faces is in choosing who will maintain the Presence, keep it fresh and ensure a certain quality standard. This person must indeed be a jack-of-all-trades; able to keep a server alive and understand the complex telecommunication issues, able to design good-looking pages to a corporate standard, able to keep abreast of the ever-changing protocols and standards, able to leap tall buildings in a single bound. And those are just the technical challenges. Politics has an even more destructive role since this person is the one who decides whether or not marketing gets that animated Graphics Interchange Format (GIF) or IS gets those printer driver downloads linked on the home page.

The role of the Webmaster should be established early on. In some cases, your marketing department (or other groups that need information on the Web site) may be savvy enough to maintain the copy, graphics, and other elements that make up the Web pages. They may even be capable of producing the actual HTML, or smart enough to have an outside design firm handle it. What, then, is left for a Webmaster to do? Maintaining the physical hardware is almost a full-time job. You should expect 7 by 24 uptime, or for those of you who prefer plain English, your server should be on-line and running all day, every day. Your Webmaster should also include a method of policing content for adherence to style. Too many sites are mishmashes of styles, looks, and feels and do not provide the single face that so many companies are trying to achieve. Your Webmaster should have final say as to what goes out on the server and therefore must also have a good working knowledge of design and style, as well as the technical ability to carry off the hardware part. Is it too much to ask of a single person?

What we are increasingly seeing is companies turning to group solutions; establishing standards committees, developing style and usage guides, creating teams of Webmasters rather than leaving it to a single person. At the extreme is Xerox's Office of the Net. Xerox has a diverse and independent organizational structure. They are primarily driven by their legendary sales staff and everyone else is overhead. So they try to make their management layer as useful in helping the salespeople as possible. The Office of the Net was established to help marry Web ideas with the sales force and to find possible ways that salespeople could use all the Net technologies in their work. What Xerox had found before the office was established was that salespeople were frustrated at the company's lack of initiative in using the Web. They needed product information, contact information—all the basics that a good business-to-business site provides. Rather than wait for Xerox corporate to launch a well-developed site, these salespeople started putting up their own home pages, devoted to their specific product line. While this is grassroots Net building at its finest, the image that went out to the customer was that Xerox was behind in its use of technology; certainly the company could not present a single face to the customer.

The office's first task was inventory taking. Before acting on any specific initiative, they simply took a look at what the salespeople were doing on their own and tried to learn some lessons from that. Their second priority was to establish strategic direction for the company as a whole. What happened was a series of summits by senior people within the organization that culminated in the direction laid out by Paul Allaire, Xerox's CEO, when he stated that Xerox was to be the "Web Document Company." The interesting thing about the Xerox approach was that it did not end in a style guide and rigid standards that each of the Xerox product lines must follow. Rather, what developed was a kind of toolbox that could be applied to many situations. The kit included boilerplate graphics that could be modified, sample HTML templates, server tools to visualize server logs, and so on. This respected the independent nature of the Xerox organization but gave each of the groups a similar starting point.

Making Decisions

Developing a first Web site is something most of you have already done, and thank God for that. What you may be faced with now is the ongoing maintenance of that site or possibly the redesigning of the site to meet some new corporate goals. What you may be looking for is the process by which decisions about the site get made. While no one approach works in all situations, we can contrast the management styles and decision making of two sites that we have already discussed in other areas: GE Plastics and UTC Carrier.

Three people made the decisions regarding the launch of the first GE Plastics site: Rick Pocock, the General Motors of Marketing Communications, Joyce Ruppert, a Marcomm specialist in charge of technical publications, and Dan Adamus, the Manager of Public Relations. That's it. Everyone else involved in the site either worked for One World Interactive (the site developers) or CommSource (the PR firm). The advantage that this team had was that decisions could be made as problems or opportunities occurred. Can't find the Ultrex design guide? Leave it out. Have more material on the various grades of Lexan? Put it in. This simple and effective style allowed the entire site to be developed and launched in less than three months from the initial meeting (where no one even knew that a Web site was being proposed) to the press conference in the Rainbow Room in Rockefeller Center.

The UTC Carrier site development was a study in corporate bureaucracy. Carrier used a standard RFP (request for proposal) process to solicit bids from venders in the early summer of 1995. This RFP was sufficiently detailed yet vague enough to provide enough rope for bidders to hang themselves. The selection committee consisted of over 20 people from the various divisions throughout Carrier and culminated in a creative presentation and technical meeting in August 1995. Once a vender was selected, the team held a three-month series of meetings to determine what the site would consist of and what the ultimate bid would include. This process resulted in a design document that detailed the components of the site. Then a contract had to be negotiated. Since the Carrier lawyers had never been on the Net, they attempted to use a standard software development contract to cover the site. While this worked for parts of the site that were clearly software development (e.g., the Oracle database that serves custom-tailored product information), it could not cover the creative areas (e.g.,

What constitutes a good graphic?). The legal maneuvering continued for over four months until a mutually agreed on contract was reached between the development team and Carrier. At this point, seven months had passed and still Carrier had no presence. The actual sourcing of materials and development of the site took three months for the bulk of the work. In fact, Carrier explicitly wrote in its contract that "time was of the essence" in completing the work. In August 1996 the site was ready for launch, except for some material that was missing from one of its divisions. Rather than wait, the site went live, one full year after the development team had been awarded the bid. But the story doesn't stop there. It was not until after the first of the year in 1997 that the final division submitted its materials and the site as outlined in the initial RFP went online. The total development effort, or the real work that went into creating the site, really only lasted three months, but the process that Carrier chose to implement its decision dragged the site development out over 15 months. (We wonder how long would that be in Web years?)

Updating

You and your Web site team worked very hard for many months; you picked the right technology; you hired a top-notch design firm; your marketers consulted with you on the flow of your catalog; you launched the site. Ready for a vacation, right? Think again. This medium demands constant updating and changing. In fact, one phenomenon that we increasingly see are out-of-date or poorly maintained Web sites. At best, these sites list some past trade shows or old press releases. At worst, the sites are exactly the same as when they were launched. What does this say about your company and it's commitment to the medium?

Part of the problem comes in when sites are designed. In the rush to get something on-line, we forget that sites need to change. Something as simple as the way we name our files becomes a major headache when the next set of Webmasters take over. Here are some basic rules of thumb to help you create an appropriate site that fits your company's needs:

- *Document Everything.* This is always the last thing that people do and the first thing that gets neglected. A file named "arg56twa97.html" may make a great deal of sense to you but it will not in three months to your successor unless you write down the thinking that went into that schema.
- *Establish an Editorial Calendar.* Your Web presence is something of a magazine, even if it is utilitarian in the extreme. Start thinking like a publisher. Establish dates and deadlines for the various areas of your site. Make sure that you have assigned editors, copywriters, and graphics people just as if you needed to get the latest issue of *Vogue* out.
- *Be Realistic with Changes.* We all start out thinking that because it is easy to change a Web page we will do so every day. There are plenty of sites on-line with huge staffs devoted solely to the maintenance of their sites and they have trouble keeping up. Don't be fooled into thinking that maintaining your site while trying to do the rest of your business will be easy. You will find other things to do. With that in mind and knowing that your site will require some maintenance, try to break up the site into sections that change at different times. For example, GE Plastics runs the Tech Tip of the Week (see Figure 8–5). This small section of the site changes, as the name implies, weekly. Other parts of the site, such as the header graphics, change monthly or quarterly to reflect changes in season or to suggest different impressions

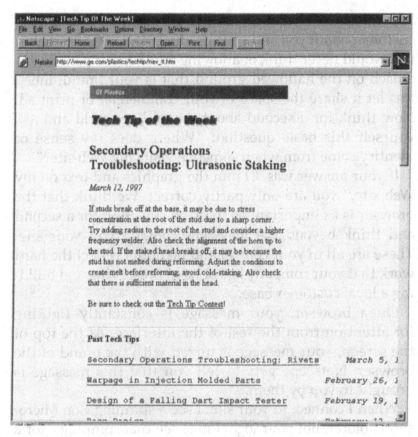

Figure 8–5. GE Plastics Ongoing "Tech Tip of the Week."

that we want to communicate to visitors. The overall effect of changing small parts of the site is a feeling that the site is constantly updating when in fact the effort required is minimal and well thought out.

Preservation—Where Does Brand Come from On-Line?

Think of all the money that your company spends on promoting its brands and perhaps even it's corporate identity.

The intention is that an undiluted message reaches the customer about you and your product. As a good marketer, you would never think of allowing another company to encroach on the hallowed ground that is your brand, much less let it share the space of your commercial or print ad. Now think for a second about the on-line world and ask yourself this basic question: "Where does my sense of identity come from when someone visits my Web site?"

If your answer was, "From the graphics and text on my Web site," you are only partly correct. We think that the browser is as important as the Web site. Stop for a second and think beyond the messages contained in your site. These are all in your control and no doubt reflect the hard work that your company has done in preserving and building a loyal customer base.

On a browser, your message is constantly fighting for attention from the rest of the interface. At the top of the screen, your message is mixed with the brand of the browser. Netscape gets to tell you that this message is brought to you by them.

When I connect to your site, I see a spinning *e* on Microsoft explorer, not your logo. I may see that spinning *e* for a good long time, staring blankly while I wait for an image to download. If I wait five minutes for a GIF to download and I only spend two minutes reading the material before I click again, whose brand just received the lion's share of my viewing time? Microsoft's, of course.

Why, you can't even keep me from choosing a bookmark or hyperlinking somewhere else. Not so great from a marketer's perspective.

A Web site is not TV, even though many would draw parallels. A television has two places for a user to interact: The television can be turned on or off, and the channel can be changed. The message, in this case visual images, conveys the sense of brand. On a typical commercial, there is one message coming from one company. My mouthwash ad does not tell me to buy a new car, nor does Sony impose anything on my screen.

So what are some solutions? Two companies have used the power of branding the browser for their own purposes: GE Plastics and AT&T. When GE Plastics launched a site in 1994, those involved realized the importance of bringing their customer base along for the ride. Very few of their large customer base had Internet access and those that did were mostly concentrated in large companies like Ford or in technology companies like Hewlett Packard. GE turned to Spry (now part of CompuServe) for help, since Spry had recently introduced their Internet-in-a-Box product. GE negotiated to sell I-Box to their customers at a rate better than they could get it at retail. But they also went a few steps further. First they replaced the Spry logo on the browser with a GE logo. Now they had instant identity—no matter where on the Net one of the GE customers surfed, that logo was always there to remind them who gave them that browser in the first place. Next GE had Spry customize the bookmarks so that they reflected a plastics industry focus. This meant work by both GE and Spry to identify any existing sites that were plastics related and program them into the software. In addition to the Spry Web browser, the I-Box also came with an Internet Newsgroup Reader. This software was also customized to immediately subscribe to those newsgroups that in some way either related to the plastics industry or heavy plastics users. And in the final stroke of marketing genius, GE insisted that the browser open to their home page. Now every time a user fired up that browser, the first thing that appeared was the folks who bring good things to life.

More recently, AT&T and its WorldNet service produced a similar result. If AT&T's marketing staff know nothing else, they know about branding. "Reach out and touch someone" ingrained the AT&T name in a few generations of phone users. Now they are turning their branding prowess on these same consumers as they attempt to become the first and largest of the ISPs to become successful in giving access to the masses. Remember the 1996 Atlanta Olympics? This was AT&T's first full-scale advertising of the service. What was notable about the commercial was the browser that was shown. To those in the know, it was a Netscape browser, but the icon spinning in the right-hand corner was clearly AT&T's. A visit to the AT&T Web site reveals a similar strategy; nowhere will you find mention of the need for anyone beyond AT&T for software, support, and even content. And the pricing cannot be beat. For AT&T long distance customers, your first five hours a month are free. No monthly charge, no restrictions, free. Should you go beyond that five hours, you are hit with a hefty $2.50 per hour but for the average consumer that five hours is enough to send some E-mail, check up on some stocks and visit a couple of Web sites. AT&T also offers the same unlimited access per month that every other cut-rate ISP in the country offers. And if it's an account that spans the globe, they just announced access in Japan. The effect of AT&T's pricing initially sent stocks plunging for other national ISPs, such as PSI, UUNet, and Netcomm. In fact, at the time that we are writing this, each of these ISPs has withdrawn from the national consumer market.

9

Getting Paid

What is more important than doing business on the Internet? Getting paid for doing business on the Internet. The reason so many people are so interested in this medium is exactly because you can make money on it. Analyst predictions vary, but most agree that a significant portion of money spent on goods and services will be spent on-line. We have seen forecasts of up to 15 percent of all consumer spending, and those figures don't count business-to-business spending or even intercompany transactions (see Chapter 7). So with billions of dollars at stake, you would think that setting up a store on the Web, accepting credit cards, conducting electronic banking, and such would be old hat by now. Think again. Many on-line users have never purchased anything on-line and don't intend to in the near term.

So what is holding back these masses from spending their hard-earned bucks at your cyber-biz? There are lots of factors, ranging from security and infrastructure to alliances between major players. But essentially, we think that the primary issue is simply lack of trust.

Microtransactions, micropayments, encryption, digital cash, debit cards, Mondex, vCard: Do any of us really understand what all of these things mean and where we fit into them? Since we generally fear what we do not understand, how can any of the companies that back these various proposals hope to build up the general trust necessary to ensure their success?

First, let's separate some wheat from the chaff. Not all electronic transactions are created equal. Credit cards are prevalent throughout most modern societies, thanks to the good marketers at American Express, Visa, and MasterCard. At its heart, a credit card is just that, credit for goods bought loaned against a future payment. The currency is hard cash, in the denomination of the local government. Digital cash, on the other hand, is an entirely new currency and one that is not issued by a sovereign nation. The question raised—that we cannot answer—is, Will people accept it?

WHO'S THE BANK?

One of the more interesting aspects of modern life is the ever-increasing gulf between what we think of as traditional banking services (checking accounts, savings accounts, home mortgages, etc.) and the people and institutions that are providing them. Cashing a check at a grocery store is more common than setting up a money market account with a securities broker. And the lines keep blurring. Very soon, Microsoft, Intuit, and a host of other computer and on-line service companies will be offering services that allow you to bank, pay bills, buy stocks, grocery shop, transfer funds, and so forth from your home computer. In fact, some of you are no doubt already doing these very things using a service like Check-Free. So the question that begs asking is the question that titles this section, Who's the Bank?

But that's a hard question. Before we can answer it, we need to cover some of the basics.

Money has become, in recent years, an increasingly abstract concept. The hard currency that you and I grew up on is being replaced by money made from bits and bytes of electricity. Credit, debit and cash cards mean that we no longer touch much of the money that we make. So in some ways the idea of digital cash isn't new, but we need to consider its implications in the on-line world.

Keeping track of the various players in the Electronic Commerce Marketplace and their various schemes is a bit like keeping track of the British Parliament. You no sooner learn who is who when the names all change. But we have to set a certain level of experience, so here goes.

SET

Secure Electronic Transactions (SET) are the universal credit card protocols, for want of a better description. Basically, both Visa and MasterCard initially had different ideas on how to process credit card transactions over the Net. Amid the hype of credit card marketing, it is easy to forget that these two bank cards really do come from different companies and are not part and parcel of the same apparatus. In any event, Visa teamed up with Microsoft and some others to make people believe that they had solved the problem, MasterCard, Netscape, and their merry band came up with their ideas on the matter and the two sides had at it. What they quickly realized was that neither of the two camps could establish themselves as the sole provider of credit card processing on the Net and that if they did not interoperate, their customers would become extremely angry and might even use, God forbid, Discover or American Express. Needless to say, when confronted with that very real possibility, these two warring factions made peace and trod down the road of enlightenment, leaving us mere mortals with, you guessed it, SET.

Commerce Servers

In traditional businesses, we take certain things for granted, like the ability to swipe a credit card and have the transaction processed, that the bank which issued the card will transfer the amount of the transaction to our bank, and that the person buying the goods with the credit card will be charged for the purchase. We take this for granted because, as merchants, we physically swipe the card, enter the purchase amount, and see a verification code. These actions are things we have done before and will do again. We place our trust in the system.

On the Net, a Commerce Server is the equivalent of the credit card swiper. Companies like Netscape, Open Market, and Microsoft all want to provide your business with an opportunity to open shop on the Web and conduct some business there. And like Verifone, the company that makes the physical credit card swipers, Netscape et al. would very much like to charge you for that transaction.

The actual schemes by which the commerce servers work vary, but the end result is the same: You can take an order and trust that your payment will arrive at your bank and your customer will be charged. Each of these companies also depends on some underlying infrastructure for their methods and these are worth a closer look.

SSL, SHTTP, and Digital Certificates

Secure Sockets Layer (SSL) and Secure HyperText Transport Protocol (SHTTP), besides having rather long and laborious technical descriptions (which we will avoid) are two of the basic methods that Commerce Servers use to provide secure connections to a Web browser. These methods both rely on encryption technology, which basically scrambles the transmissions up and then descrambles them later (sorry to get so technical). Across these secure connections, the very being of Electronic Commerce can fly, ensured and endowed by technology that it will safely reach its appointed destination.

Like SET, SHTTP, and SSL came from differing groups and both entered the spotlight at around the same time in the spring of 1995. SHTTP was proposed by CommerceNet, the consortia that we mentioned back in Chapter 3, as well as by Spry and Spyglass, who at the time were leading browser manufacturers. SSL was thought up by the folks at Netscape. Realizing that a split in standards would only hurt both groups, Netscape, Spry, and Spyglass along with IBM and Apple decided to fund a new company, Teresa Systems that would develop a common set of standards that worked together. This allowed companies to adopt either SHTTP or SSL and be assured that browsers and secure servers would support both. Teresa systems is also noted for developing the vCard system, a method of exchanging electronic business card information.

But who is to say that the person who contacts you, as a merchant, is indeed the person responsible for paying you? In a retail setting, it is easy to see someone hand you a wad of cash. If they pay by check you can ask for a driver's license to verify identity. With a credit card, you can check a signature, or if you are a bank, ask for the last four digits of a Social Security number. But what if the only contact between your store and the buyer is electronic? One person can sit at another's computer and use a Web browser. A person may even have more than one Internet Access account, more than one E-mail account, more than one on-line identity. How then do we separate the proverbial wheat from the chaff?

That is the role of the digital certificate. A digital certificate authenticates us to the Commerce Server, letting it know that we have not (or have) stolen a credit card and are on a buying spree. These certificates are issued from a standards body, like a bank, local government, or a corporation and filed with RSA, the cryptography people. When a certificate is used, say to send E-mail to your bank requesting an account balance, the bank would check the stamp, see that your certificate was verified by RSA, and make the leap of faith that the E-mail came from you. Is there trust involved? You bet. But remember, trust is what the big hurdle is.

Cryptography

The concept of cryptography is fairly simple. Transpose something for something else using some algorithm, and you have a message that is unintelligible except to those who know what the algorithm is. For example, the alphabet could be represented using the numbers 1 through 26 where 1 equals A and 26 equals Z. This becomes 20–8–1–19. This particular algorithm would be called a two-bit algorithm since it uses up to two numbers to represent a single letter. Simple to understand and simple to decode, even to non-cryptographers. But suppose I used 8 numbers to represent each letter, or better yet, 64 numbers to represent each letter. Suddenly a person can no longer break the code and a computer is required. And a very big computer at that. In fact, a 64-bit encryption scheme is damn near impossible to break and I only say nearly impossible because someone might someday invent a computer big enough to handle it, but no computer exists today.

E-Cash

E-cash is the electronic equivalent of a traveler's check. You send a participating bank a dollar of currency (in whatever denomination) and they send you back the equivalent amount of E-cash. The advantage (and this is where E-cash differs from some other methods, including traveler's checks) is that E-cash is anonymous. No record is kept by the merchant of who submitted the payment, only that the payment was good. This, incidentally is true of any minted currency. E-Cash is a product of DigiCash.

Mondex and SmartCards

Once we agree that some sort of digital cash is a good thing, the next question that comes up is, How do I carry it? It is a good question, indeed, since many scenarios rely on you using the cash on a computer.

Mondex is electronic cash on a card. The card has a microchip that holds balance information as well as a means of providing security. Balance readers are available so you know how much you have on the card. Specially equipped cash registers, phones, and other devices, not unlike a credit card swiper, are available to accept payment. The difference between Mondex and other SmartCards, a credit card, or a check is that the value of the transaction is immediately transferred between the parties involved in the transaction. No more waiting three days for Visa to deposit your funds (if you are a merchant), but it also means no more float on a check. SmartCards also mean an end to the old saying, But I must have money because I still have checks left.

CyberCash

One of the founders of CyberCash, Dan Lynch, co-wrote a very good book for people interested in the topic of Electronic Commerce called *Digital Money*.[1] The focus of Cyber-Cash is to provide a secure Internet transaction, regardless of whether that transaction is cash, check, or credit card. Without getting too deep into the technology, CyberCash acts as a clearinghouse for a banking network, providing a single point of contact for transactions. For example, if you were to pay John Wiley & Sons (the publisher) for this book on the Web using CyberCash and your credit card, Cyber-Cash would encrypt the transaction, provide verification services for Wiley, provide Citibank a way to debit your MasterCard account, and provide you with the overall method to pay for the book. They have also introduced CyberCoins as a micropayment method that leads us to another question.

WHAT GOOD IS IT ANYWAY?

The smallest denomination of a currency sets the lower limit for what a merchant can charge for a particular good. And it also defines the smallest unit that a merchant is willing to produce or form into groups of units. In other words, you can't sell something for less than someone can spend. This does not work at the high end of the currency scale because you can always put denominations together to create greater amounts.

That is why people sell a box of paper clips rather than individual ones or books instead of sentences. But what if you only need a single paper clip? Your government, by virtue of restricting the lowest level of spending, is placing an unfair and unjust burden on you to buy more than you need! Jack Sprat beware!

Digital cash, by the very virtue of its digitalness, breaks that tyrannical government's back by allowing you to spend as little as you like. Want a paper clip? That will be .012 percent of a penny. Care for a well-crafted sentence? 1/8 of a cent please. Since the smallest denomination is no longer pegged to a physical representation, the lower limit is defined only by the merchant.

Think about the last trip you took abroad. Remember how much fun it was to exchange money into another country's currency. Remember how good you felt when the bank exacted their toll for the exchange? Your dollar was weak against the yen? Too bad. Your Deutsche mark was strong against the pound? Good for you. And then you had the distinct pleasure of exchanging it all back when you returned home.

Digital cash knows no boundaries. Swipe your card in the localization swiper at your digital cash friendly airport and you are ready to spend. Afraid that the dollar is about to plunge? Get paid in Swiss francs. Care to arbitrage the pound? Get that refund in pesos. Your digital account does not care. Currency becomes fluid and global.

The value in microtransactions is literally in the eye of the beholder. Think briefly about the time you spend watching television. Assume, that on a typical night you watch TV for two hours of standard network broadcasting and cable fare. And let us suppose that you also receive four movie channels that add to your monthly cable access bill. If you are at all like us, during that two hours you watch more than the four half-hour shows slotted for that time slot. Channel surfing is a birthright and the reason God made remote controls. In fact, some of us have been known wrongly to use the argument that channel surfing is the only way to justify the added expense of cable or a satellite dish. The key word in that last sentence was "wrongly" because you are charged for the channel whether you watch it or not. And the companies that advertise on those channels are charged for their ads whether you watch them or not.

What if you could pay only for the television that you watched? And what if the advertisers could pay only for the people that had their televisions tuned to the programs that ran their ads? Watch an hour of the Jets game? No charge. Watch three minutes of PBS? One dollar. Sit through an ad for life insurance? MetLife pays ABC 50 cents. That, in a nutshell, is the idea of a microtransaction.

WHO IS "THE BANK?" REVISITED

We started this discussion with the question, Who's the Bank? and we return to it now with a bit more understanding about the problem.

What we have underscored is that the services traditionally provided by a bank are being increasingly provided by someone else.

First Virtual, the Internet Bank

Old-school savings and loans were a pretty simple affair. You set up an account and deposited your money, which you withdrew as you needed it. The bank made its money from loans, float, and transaction charges. First Virtual operates on the same basic philosophy. Depicted in Figure 9–1, the First Virtual system is a model in simplicity and has been set up and running as the longest secure payment system on the Net (Figure 9–1). While other methods require elaborate encryption techniques and specialized software, First Virtual relies on tried and true E-mail.

First Virtual got its start in October 1994, when Spry, GE, Netscape, Times Pathfinder, and a host of other businesses

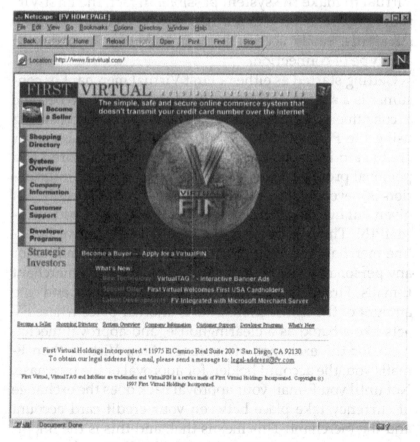

Figure 9–1. First Virtual Web Site.

woke up to the potential of the Web. The goal of the founders was to create a payment system that was above all else safe, but also one that would not require additional software to be installed by either the merchant or the consumer. By virtue of its simplicity, the system would also be low-cost and simple to implement. Unlike other solutions, First Virtual did not need to spend millions on software development, never had to lobby the government to allow it to distribute the payment system around the globe, and certainly did not need large banks and credit card companies to strategically align with First Virtual to make the system possible. In fact, the First Virtual concept is very much like the Net; based on open standards, easily accessible, and usable by anyone with any type of connection.

Getting started as either a First Virtual merchant or consumer is a simple and straightforward process. To become a consumer, or someone who wants to buy something using the FV method, you set up an account using your credit card instead of cash. This information is stored in a personal profile by First Virtual, your credit card information is never given out to anyone else and certainly not given out over the Net. First Virtual then issues you a VirtualPIN. The PIN is what you, the consumer, give to online merchants instead of your credit card information or any personal data. When you place an order, the merchant E-mails First Virtual with your PIN number and the amount of the transaction. This is another place where FV acts like a bank, as a clearinghouse, and approval mechanism for the exchange of currency. First Virtual then E-mails you, the account holder, for approval of the purchase. Not until you E-mail your approval back does the exchange of currency take place between your credit card account and the merchant. This means that, and this is the important part, no information that could fall into the wrong hands ever gets transmitted on the Internet. The only information sent between you and the Merchant and the Merchant and First Virtual is a VirtualPIN and a purchase amount. No E-mail address of who is doing the purchasing, since FV stores that information when you set up an account. No description of you, again this is stored with First Virtual. Your credit card number is stored with your profile so that never crosses the Net. This significantly lowers the barriers to entry for people looking to buy over the Net. No secure browsers, no secure servers, no encryption of any kind is ever used to place the transaction.

But what does it cost the merchant? You need an E-mail account, since that is the communication method. And you need a bank that accepts direct deposits via a wire transfer. The fee structure is incredibly low and affordable. It includes an annual fee of $10.00, $0.29 per transaction, plus 2 percent of the total sale and $1 for payment processing and the direct deposit. The 2 percent fee is based on what First Virtual pays to process your credit card.

Strategically, this low-cost model makes a great deal of sense. The cost to the consumer is virtually free ($2.00), so there is no barrier to entry for anyone who uses E-mail (remember, this isn't Web dependent). Merchants basically pay what MasterCard or Visa would charge them (2% transaction fee) so there is no barrier for merchants. The drawback? Lack of sizzle and PR value. At a time when Netscape, Microsoft, IBM, MasterCard, AmEx, the major banks, and probably even our grandmothers are announcing elaborate plans that include sophisticated schemes and will cost millions to implement, completing basic transactions via E-mail just doesn't sound cool enough. The Internet-related trade magazines do not devote time to FV, because there isn't a story to tell. No one has had their credit card information stolen, no one has hacked their server, just boring old business on the Internet. E-mailing PIN numbers is too simple a solution in this complex arena, but in the end, when Main Street becomes Cyber-Main Street, good old E-mail may be just what smaller shops want.

The Net as ATM

SmartMoney cards are coming, in fact, parts of Europe already use them regularly. But what are they? Think of them as a credit card that uses digital money as the exchange mechanism, but instead of getting a credit (or taking out a 30-day loan as it were), you treat the transaction like cash. This is not a debit card, that relies on an account in a bank somewhere to draw from. It is more like a prepaid calling card, where the card holds a certain denomination and draws against that amount until you have spent your limit. When your limit is reached you can update the card to new amounts. This is accomplished by either inserting your card into an ATM-like machine and instructing it to debit from an existing bank account or by inserting money into a slot and having the value transferred to the card.

The next logical extension to the SmartMoney concept is to build hardware into your PC that allows you to instantly update your card with a new balance. Think of the convenience. No more waiting around the ATM, worrying that someone will be looking over your shoulder for your PIN code. No more worries about going to an ATM late at night and getting mugged. No more machines that run out of 20-dollar bills and so you get stuck with 200 dollars in 5-dollar bills.

The underlying infrastructure that would make a scenario like that possible is the same sort of infrastructure that would be required for any electronic transaction: secure encryption and access to an electronic clearinghouse. Getting hardware manufacturers to include the device in their PC designs would be as simple as giving them a percentage of the transaction fee. If you were Compaq and could receive one hundredth of a percent of every cash transaction that took place on the millions of computers that you sold would you think very long about whether or not to include that piece of hardware? Of course not.

So far we have dealt primarily with software and network designs as the principal driving factors toward secure, safe, and efficient electronic commerce. There are, however, some folks that think a chip inside your computer is the real key to get this commerce ball rolling. What chip, as it were, through yonder window breaks? *Wave Systems.*

Wave Systems was founded by Peter Sprague in 1988 to provide hardware for secure, real-time, on-line transaction processing. The key to their approach is a hardware metering system, WaveMeter. WaveMeter allows for usage-based content pricing. If you think of the way a municipality charges for water, then you will understand the basics how WaveMeter works.

In order to work, the WaveMeter needs to reside either on your personal computer or on a network server. WaveMeter communicates with the WaveNet transaction processing network to provide account information, billing, credit card processing, and fund distribution.

The idea is that since this is hardware and not software, the encryption methods can be hard-wired into the chip thereby providing a higher level of security than a software-based method. Can hackers crack the chip? Only time on the Net will tell.

THE MOVEMENT WITHIN

When we think about electronic commerce, digital money, and the like, we tend to think about how it affects the consumer marketplace. The articles in major trade publications and certainly in the general business press deal almost exclusively with the idea of consumers buying something off the Internet. For many of you, this is in fact the appeal of providing a Web site in the first place.

But it is time once again to put on our strategy caps and think about where the real revolution in electronic commerce will take place. If your cap is on straight, then you have already thought that it is the business-to-business market; maybe even a few of your thoughts have come out, It is inside our company! To you folks, we heap on much praise because it is both the business-to-business market and the intranet where the largest impact will take place. Why? Think of it in terms of transaction costs and the final cost of the good sold. If you buy a sweater from a mail-order outlet, the transaction price of the store taking your money in exchange for the sweater is reflected in the price of the sweater. One transaction, one price. But in making the sweater, the company that sold it to you had to contract it to some Third World sweatshop, ship the sweater to a distribution point, warehouse the sweater, and ship the sweater to the retail outlet before finally making the sale to you. In accounting for the cost of the good, each of these operations takes on a certain transaction cost. Some of the cost of each transaction is tied to the flow of information from the purchasing party to the receiving party. Purchase orders, invoices, checks, and such all increase the cost of the good, in turn either reducing the margin to the reseller or increasing the cost to the consumer.

For years EDI, or electronic data interchange, has been hoisted up the flagpole as the solution to all problems regarding the exchange of information between buyers and sellers. Need to exchange purchase orders. No problem, we'll use an X5698 protocol. Need to exchange money? No problem, we'll use the good old X500879 protocol. Everyone could agree on the protocols, everyone could agree that it was a good thing. The problem? The costs involved in maintaining an EDI network prohibited smaller businesses from joining. Indeed, the complexity was prohibitive even for many midsize firms. In addition, the networks relied on expensive leased lines, and the EDI network carriers gouged their customers (well, maybe not gouged, but certainly made some money).

Then along came the Net. Now we can take the wonderful promises of EDI and apply them regardless of the size of the business. Why? Because the marginal cost of the transaction, stated as the cost of transmitting the information between parties, is reduced to near zero. Once again, the idea of a zero transaction cost surfaces, but it means that more businesses can reasonably afford to conduct electronic transfers of money. And because the EDI standards are global and the Internet is global, small businesses can conduct their affairs anywhere. The effect will be that over the next few years, more businesses in more industries will be connecting to the Net for the sole purpose of transmitting EDI information.

SETTING UP SHOP

Perhaps the most perplexing question in the world of electronic commerce is the most basic, When does my company start? If there were a simple answer to this simple question, this chapter would be a paragraph and the book much shorter. First let's assume that you are looking to do full-blown electronic commerce; EDI, credit card verification, digital certificates, the works. There are two basic paths to take and borrowing from Robert Frost here is where the road diverges in the yellow wood. The road more traveled will be partnering with a service provider. The road less traveled will be doing it yourself. We'll start with the road most traveled by, not that it makes a difference.

Partnering

The larger Internet Service Providers (ISPs) will soon be offering soup-to-nuts electronic commerce solutions. These will consist of an electronic catalog builder, secure transaction capability, registration for digital certificates, and hosting of your site. Right now, you pay a premium to receive this kind of service because of the newness of selling on the Net. When you open a physical store in a shopping mall, you expect certain things from the landlord: heat, electricity, lighting, parking, and so on. So you should expect these cyber-landlords to include some of these same things. It will be in the best interest of the ISPs to make this as easy for you to participate in as possible, since the barrier to entry on your part is the perceived work it involves. After a relatively high initial offering price, we expect the price to drop dramatically as the providers iron out the wrinkles in providing the service. The reason? The profit opportunity for them is not in helping you create the catalog but in the percentage of the transaction. For a company like AT&T that has the bandwidth to host thousands of sites around the globe, the profit potential is enormous. Expect to see this service given away at some point in the near future.

The smaller ISPs will offer a similar service but at a steeper cost. Their selling point will be a more direct and customized service. They won't have the aggregate economy of scale that an MCI has, but they should be able to remain in the game if they move to more hand-holding of customers. There is a big advantage to being able to provide house calls and face-to-face service. Expect the better of the small providers to grow into small regional providers and expect the very small providers to become even more focused.

Do-It-Yourself

The road that will be less traveled, to return to our poetic metaphor, will be the do-it-yourself method. All the pieces of the puzzle will be available either as freeware or through the existing server venders (Microsoft and Netscape to name but two). The main advantage to setting it up yourself is the reduction in marginal cost. Going through a service provider means a percentage of each sale not only goes to the credit card clearinghouse (or the digital cash clearinghouse) but also to the ISP that is providing the service. If you intend to run a high-volume, high-transaction shop, this cost might outweigh the reduction in marginal costs in producing a paper-based catalog. We expect that the ISPs will offer a service that will be substantially lower than paper to entice the switch from paper, particularly to enter, but not so low as to negate a very high profit margin for themselves. The strategic reasoning to follow is to maintain price control and system control, and remove yourself from the cookie-cutter approach to catalog building that the ISPs will employ.

Stop Thief!—The Security Issue

If Dorothy's tornado were to land her in CyberOz, she might exclaim, "Crackers and Hackers and Geeks, Oh My!" And like Dorothy in Oz, many of your fears surrounding security on the Net are as unfounded. Others come from directions you might not expect, not unlike the ambushes of the Wicked Witch or a Flying Monkey.

Proponents of the encryption and the secure exchange of moneys across the Net love to point out that talking on a cordless or cellular phone is less secure than the Net and yet people willingly give out sensitive information on phones every day. Who among us, besides conspiracy nuts, thinks twice about phoning in to a mail-order house and ordering the latest in chocolate macadamia nut coffee. For the most part, this argument is correct. A simple and easily purchased police band scanner can listen in on cordless and cellular phone calls, and if persistent, a determined thief can pirate credit card numbers, Social Security numbers, and so on.

What the argument fails to consider is that intercepting credit card information from cellular phone calls is a relatively random event. Think of the endless calls that a thief would need to intercept, listen in on, determine whether there was to be an exchange of credit card numbers, then listen for the numbers, and other information. Time consuming to say the least.

A *cracker* (brief definition—a cracker is an evil hacker, who breaks into another's computer for the purpose of wrongdoing; a *hacker* is just about every computer geek under the sun, and they know who they are), on the other hand, needs only to write a program that sits on a commerce server, scanning through messages for a sequence of numbers that looks like a credit card number. These numbers are conveniently written to a temporary file which the cracker picks up before heading for a small Caribbean island or until the Feds kick in his door. These programs are relatively easy to write but hard to implement. Cracking a computer is no trivial task, but a determined individual usually does not crack just one. And a very good crack may go undetected for days, weeks, or months, or even forever since good security is not something that every computer connected to the Net has. In fact, a break-in happens to just about everyone.

General Electric, one of the first on the Web, did lots of groundbreaking work and had an R&D team that pioneered 3D visualization across the Net, and so on. But GE learned that the Net can be a not-so-nice place during Thanksgiving weekend 1994, not too very long after they put their first Web server in place. A cracker broke into the GE R&D server through a Web form that was improperly coded, allowing an entry of any length and to contain any string of commands. Once inside, the cracker quickly gathered account information found on the computer (login names and passwords of lots of employees). In fact, there was quite a bounty because this particular server also acted as the E-mail gateway for GE Corporate R&D.

The break-in was detected on another machine within the GE internal network the evening before Thanksgiving. It seems the cracker was very familiar with hiding his steps on a version of the Unix operating system from Sun Microsystems called Solaris, but had less experience with the version of Unix sold by Digital Equipment. And therein lay his downfall. An administrator found the crack on the Digital machine and quickly traced the route back to the Sun Web Server in GE Corporate R&D. Within the span of several very long days, a team of computer security experts were able to trace the entire path of the cracker, remove his files from their network and plug the hole in their internal security systems. This was, as you might imagine, a fairly massive undertaking for a company the size of GE, especially since their internal network is international in scope.

The general feeling on the Net is that you can expect some sort of attack to occur on your system at some point. Many of these attacks will be successful, and your equipment will be compromised. While that may seem alarming, what happens in a majority of break-ins is nothing at all. Sometimes the attacks are merely done in the Hillarian spirit of "Because it's there." If the attack is more serious, the thieves will want to hide their steps by jumping to and from as many computer networks as possible. The more computers in the path, the more complex the trail for the authorities to follow. So your break-in may be nothing more than an attempt to cover up a path, a sort of backward walk to erase footsteps the crackers have left behind.

10

The Re-Wired Business

Until this point, we've discussed what Internetworking and intranetworking can do for specific businesses; examples include retailing through Web sites, setting up electronic markets, reengineering business relationships, moving information around the organization; and we suggested some key strategies for moving ahead in this new electronic world. For many businesses, however, what will be most important is not the microlevel application of technology, but the macrolevel understanding of fundamental changes in the industry in which they compete.

Information-based business is now undergoing massive change. In some cases, these changes started before the advent of Internetworking, but the Internet has accelerated, and will continue to accelerate, these changes. In other cases, the Internet is providing new business and growth opportunities that are allowing new entrants into a market or changing the relationship between the present players. In this chapter, we will look at how business and industry are being transformed due to their exposure to the Internet and the Web.

NET-ACCELERATED CHANGES

What are the fundamental changes that the Net and Web are pointing to? As discussed throughout this book, we view the Net as an *accelerator*, not necessarily an instigator. Many of the fundamental changes in business evident in 1990, including moves toward mass customization, service orientation, and rapid distribution, are added by electronic networks. But the ideas existed before millions had heard of the Internet.

Table 10–1 shows some of what we call Net-accelerated changes—things that are happening faster. They range from the pretty obvious ("new entrants") to some wilder ideas ("process leasing").

It is clear, at least to us, that the Net has not resulted in a rapid move toward core functions *or* diversification. Consider newspaper publishing, for example, where a number of notable newspapers have used the Web to diversify (*The Electronic Telegraph* in the U.K. moving into value-added services such as fantasy football, the *Albany Times-Union* in upstate New York becoming a service provider), disaggregation will allow many small content providers to be just that—content providers. They can distribute work electronically, either to physical outlets or straight to consumers. The structure of electronic publishing will be more complex than physical publishing. Likewise, a small financial analysis company will be able

Table 10–1. Some changes in Net-centric businesses.

Net-Accelerated Changes	Explanation
Retreat or diversification	Some business will use the virtual infrastructure to retreat into core business; others will use it to diversify.
New entrants	Due to low entry barriers, businesses will be able to enter new markets.
More start-ups, more shutdowns	Due to low start-up costs, entrepreneurial activity will continue to be intense.
Accelerated globalization	Services will be global, even if the business does not anticipate this.
Electronic alliances	Businesses will use the electronic world to quickly develop and test alliances.
From products to services	Items we view as products (such as greeting cards) will become electronic services.
The beta-test model	Services will be released in beta-test mode so as to better understand customers and facilitate early entrance.
Process leasing	Business and customers will lease processes available electronically rather than implement them in their organization.

to provide and sell services in multiple contexts to targeted clients and browsing customers.

Entry into new markets is already apparent. What business domain can not be entered given the pervasiveness of this medium? The moral: Those that understand the electronic world can plot a strategy based on IT and the Net. Witness: Microsoft's entrance into many businesses that

262 Re-Wiring Business

span several industries. Those that can't will see others using networking to chip away at their business.

We can't keep track of the number of Net-based businesses that have started since early 1994. Low barriers to entry and small start-up costs have generated a slew of Web-based retailers, financial analysts, magazines, and tarot card readers. Although the combined revenue of these is, as discussed earlier, less than a good week for Wal-Mart, the long-term impact may be considerable. Who is willing to bet against one or two Fortune 500 companies in the year 2020 being 1990's Net-based start-ups? Wall Street appears to be betting that a few will be.

Accelerated globalization is a pretty safe prediction; we don't feel too clever predicting the extension of an already undeniable change. But what is interesting is the extent to which a business will be dragged, in many cases, into unintentional global competition. Newspapers now on the Web find they have readers from all over the world: Why read a local report of, say, a soccer game when you can get a report from where the game took place (language allowing)? Why rely on Asian news filtered through the cultural predilections of a local national paper when English language versions of Asian newspapers are on the Web? (The South China Post now has a global readership.)

The opportunities for alliances based on electronic commerce are immense. Why, when reading a financial newspaper, is it not possible to click through the name of a stock and go straight to a screen that allows purchase? Why, when visiting a start-up such as ESI that provides trading facilities, is it not possible to click through to multiple analyses? The answer is simply that these alliances are not yet in place, but consumer demand will accelerate their creation. Product review and information will become indistinguishable from the purchase ritual. This is already evident on some Web book and CD stores, where inverted catalogs (see Chapter 5) allow browsers to read reviews, sample, and then purchase.

This movement toward marketing based on information rather than product attributes, such as cost, will be accelerated. Ultimately, retailers may be differentiated based on the services they provide (information, product reviews, after-sales support, etc.) rather than product display and delivery, much as now happens in the business-to-business environment. Further, some information-loaded physical items such as greeting cards, financial prospectuses, and food recipes can be delivered electronically. Consumer desire to find a provider of the physical item will be replaced by expectation of finding a service provider for the necessary information. Forget about who sells the best greeting cards: which sites provide the best design facilities for unique electronic cards?

As the move toward a value-added and service-based electronic economy gathers momentum, the beta-test notion will be more prevalent. Netscape changed the way software is sold, possibly forever, by just putting their beta-test versions on a server. You download it, play with it, buy it when it becomes the current version. You want to release a new electronic newspaper? Do the same thing: early issues for free, get customer feedback. Since the cost of delivery over the electronic infrastructure is so low, beta-testing costs little.

Our final idea is one that takes the notion of Net-based services to the extreme. If the cost of delivery is virtually zero to both supplier and customer, why will businesses continue to perform processes in-house when they can be seamlessly integrated even if provided elsewhere? Rather than hire accountants, why not define accounting processes and outsource them? This happens now, but in a very restricted sense, when outsourcers recreate the internal process and manage it. The next step is for accountants, lawyers, and others to define generic processes that are available over the Net. These are then purchased from the electronic marketplace. The result may be that the 21st-century business will have the majority of its processes performed extraneously: The location of these processes will be irrelevant, and all electronic integration will be seamless.

FINALLY

Here's another scenario, similar to those that started the book.

A semiretired banker maintains a portfolio of invest-ments. These provide some income, but are mainly there for when full retirement beckons. To aid in managing this port-folio, he purchases specific analyses when appropriate from financial consultants. Some analysis consultants will also directly contact him when they think they have an analysis they can sell. Stocks and bonds are traded at will via a bro-kerage service. For cash, multiple accounts in differing cur-rencies are maintained, and money is transferred between them as desired. He monitors his portfolio to check that cer-tain portfolio bounds are not exceeded, thus providing some simple risk management.

Accounting for all operations is provided via an external service that monitors all transactions, producing profit and loss accounts. At tax time, tax preparation work is sent to a market for bidding. The winner of the bid prepares taxes for all necessary governments and then submits them. If audited, the accounting firm used can provide val-idated copies of all monitored transactions.

Pretty much everything identified in the previous two paragraphs can be performed electronically from the com-fort of a home PC. Our banker need never leave his study.

Electronic consumer brokerage, Web-based portfolio analysis, E-mail alerting services, home banking in multiple countries, accounting via a third party, Web-based tax preparers (who will quote for work), and stock monitoring software all exist. Piecing them together and using them is not trivial: In 1997 our home-based banker also has to be PC and Net savvy. But in 2000, piecing the bits together may be trivial. The individual consumer, the home-based business, and the Fortune 500 company will all have access to the best information, processes, and services.

Multiply this across all areas of information-based services, throw in the value-added information component of many products, and imagine the business environment 10 years from now. The virtual economy is and will be increasingly competitive, global, and fast-paced in ways that we cannot yet comprehend. We hope that the ideas and views expressed in this book will help you rethink your business strategies and style. If you can embrace and infuse the Net's many opportunities in your own departments, team, and organizations we think you will be richly rewarded. Start re-wiring now!

Notes

Introduction:

1. Ives, B., and S. Javenpaa, The global network organization of the future: Information management opportunities and challenges, *Journal of Management Information Systems*, 10:4, 1994, pp. 25–57.

Chapter 1

1. Sterne, J., *World Wide Web Marketing*, New York: John Wiley & Sons, 1995.

Chapter 2

1. CommerceNet/Nielsen Internet Demographics Survey, 1995, available at http://www.nielsenmedia.com/whatsnew/execsum2.htm.
2. Hoffman, D.L., W.D. Kalsbeek, and T.P. Novak, Internet and Web use in the U.S., *Communications of the ACM*, 39:12, 1996, pp. 3466–4108.
3. Keen, P., *Shaping the Future: Business Design through Information Technology*, Cambridge, MA: Harvard Business School Press, 1991.

4. Open Market maintains a directory of all commercial Web sites in the United States, and list the total number. Since many of these sites host pages for multiple businesses, this number is in fact an underestimate.

5. Rayport, J.F., and J.J. Sviokla, Managing in the marketspace, *Harvard Business Review*, November–December 1994, pp. 141–150.

Chapter 3

1. "From the Ether Metcalfe's Law: A network becomes more valuable as it reaches more users," by Bob Metcalfe, October 2, 1995, *Infoworld*, (Vol. 17, Issue 40).

2. Dave Winer, "If the Net were Smarter," March 8, 1997, *DaveNet*.

Chapter 4

1. Rick Levine, Sun's *Guide to Web Style*. Copyright 1995. http://www.sun.com/styleguide/tables/Welcome.html.

Chapter 9

1. Lynch, Dan, *Digital Money*, New York: John Wiley & Sons, 1995.

References

Aguilar, F. J., General Managers in Action, Policies & Strategies, Oxford University Press, 1992.

Anderson, H., Why the Internet chews up business models, *Upside*, August 1995, available at http://upside.master.com/print/aug95/9508f1.html.

Applegate, L. M., and J. Gogan, Paving the information superhighway: Introduction to the Internet, Teaching note #9-196-006, *Harvard Business School*, August 1995.

Benjamin, R., and R. Wigand, Electronic markets and virtual value chains on the information superhighway, *Sloan Management Review*, Winter 1995, pp. 62–72.

Berlingeri, I., S. Brett, and E. Gokyigit, Cultural Imperialism and the Internet, Harvard Law School student papers, May 1996, available at http://roscoe.law.harvard.edu/courses/techseminar96/course/sessions/culturalimperialism/intro.html.

Bernard, R., *The Corporate Intranet*, New York: John Wiley & Sons, 1996.

Bloch, M., Y. Pigneur, and A. Segev, On the road of electronic commerce—a business value framework, gaining competitive advantage and some research issues, Technical Report, Fisher Center for Information Technology, University of California, Berkeley, CA, 1995, available at http://haas.berkeley.edu/~bloch/docs /roadtoec/ec.htm.

Clemons, E. K., and B. W. Weber, London's big bang: A case study of information technology—competitive impact and organizational change, *Journal of Management Information Systems*, 6:4, 1990, pp. 41–60.

CommerceNet/Nielsen Internet Demographics Survey, 1995, available at http://www.nielsenmedia.com/whatsnew/execsum2.htm.

Computer Mediated Communication, special issue on Web Business Models, available at http://www.december.com/cmc/mag/1996 /jun/toc.html.

Computer World, Web payoffs now, November 20, 1995, pp. 90–94.

Computer World, On-line to the rescue, December 4, 1995, p. 147.

Ellsworth, J. H., and M. V. Ellsworth, *Marketing on the Internet*, New York: John Wiley & Sons, 1995.

Emery, V., *How to Grow Your Business on the Internet*, Scottsdale, AZ: Coriolis Books, 1995.

Financial Times, How to join the on-line revolution, March 1, 1995.

Gatignon, H., B. Weitz, and P. Bansal, Brand introduction strategies and competitive environments, *Journal of Marketing Research*, 27, 1990, pp. 390–401.

Gurbaxani, V., and S. Whang, The impact of information systems on organizations and markets, *Communications of the ACM*, 34:1, 1991, pp. 59–73.

Gurley, J. William, Defending Amazon: As if It's Necessary, *Above the Crowd*, Deutsche Morgan Grenfell Technology Group, Newsletter Issue 97-01.

Hoffman, D., T. P. Novak, and P. Chatterjee, Commercial scenarios for the Web: Opportunities and challenges, *Journal of Computer Mediated Communication*, 1:3, 1995, available at http://shum.cc.huji.ac.il /jcmc/vol1/issue3/hoffman.html.

Hoffman, D. L., W. D. Kalsbeek, and T. P. Novak, Internet and Web use in the U.S., *Communications of the ACM*, 39:12, 1996, pp. 3466–108.

Hopper, J., Editorial standards for Web advertising, available at http://www.jimhopper.com/ads-pre.html.

The Independent, Retailers are trying to hold back the Internet tide, August 23, 1996, p. 19.

Ives, B., and S. Javenpaa, The global network organization of the future: Information management opportunities and challenges, *Journal of Management Information Systems*, 10:4, 1994, pp. 25–57.

Johnson, S., Retail systems: No longer business as usual, *Journal of Systems Management*, August 1992, pp. 8–35.

Johnston, R. H., and M. Vitale, Creating competitive advantage with interorganizational information systems, *MIS Quarterly*, 1988, pp. 153–165.

Keen, P., *Shaping the Future: Business Design through Information Technology*, Cambridge, MA: Harvard Business School Press, 1991.

Klein, S., The configuration of inter-organizational relations, *European Journal of Information Systems*, 5, 1996, pp. 92–102.

Konsynski, B., A. Warbelow, and J. Kokuryo, Aucnet: TV auction network system, Harvard Business School Case 9-90-001, Cambridge, MA, 1989.

Lynch, D. C., and L. Lundquist, *Digital Money*, New York: John Wiley & Sons, 1996.

Malone, T. W., J. Yates, and R. I. Benjamin, Electronic markets and electronic hierarchies, *Communications of the ACM, 30*, 1987, pp. 17–21.

Mitchell, W. J., *City of Bits*, MIT Press, 1995.

Modahl, M. A., and S. H. Eicher, The Internet economy, *Forrester Report: People & Technology Strategies*, 2:5, 1995, available at http://www.forrester.com.

Newsweek, This Web's for you, April 1, 1996, pp. 74–76.

Novak, J., and P. Markiewicz, Setting up shop: The Kaleidospace experience, *Internet World*, January 1995, pp. 67–72.

O'Keefe, R. M., Marketing and retail on the World Wide Web: The new gold rush, in *Special Management Report on Electronic Commerce*, Nikkei Publications, Japan (in Japanese), 1995, pp. 58–65.

Pitkow, J. E., and C. M. Kehoe, Emerging trends in the WWW user population, *Communications of the ACM, 39*:6, 1996, pp. 106–108.

Press, L., Commercialization of the Internet, *Communications of the ACM, 37*:11, 1994, pp. 17–21.

Quelch, J. A., and L. R. Klein, The Internet and international marketing, *Sloan Management Review*, Spring 1996, pp. 60–75.

Rayport, J. F., and J. J. Sviokla, Managing in the marketspace, *Harvard Business Review*, November–December 1994, pp. 141–150.

Schmid, B., Electronic markets in tourism, in W. Schertler et al. (Eds.), *Information and Communications Technology in Tourism*, New York: Springer, pp. 1–9.

Solomon, B., TV shopping comes of age, *Management Review*, September 1994, pp. 22–26.

Sterne, J., *World Wide Web Marketing*, New York: John Wiley & Sons, 1995.

Tenegra Award for Internet Marketing available at http://www .tenegra.com.

U.S.A. Today, Promise of the Internet, November 13, 1995 (Special section).

U.S. Bureau of the Census, *Statistical Abstract of the United States*, 119th Edition, 1994.

U.S. News and World Report, Gold Rush in Cyberspace, November 13, 1995, pp. 72–83.

van Heck, E., E. van Damme, J. Kleijnen, and P. Ribbers, New entrants and the role of information technology: The teleflower auction in the Netherlands, Technical Report, Tilburg University, The Netherlands, 1996. See also the business case at http://kambil.stern.nyu.edu/teaching/cases/auction /flowers.html.

Webster, J., Networks of collaboration or conflict? Electronic data interchange and power in the supply chain, *Journal of Strategic Information Systems*, 4:1, 1995, pp. 31–42.

Index